29.95

"... and there is reward for life in water and field."

—Paul Horgan

Great River: The Rio Grande in North American History

The Chile Chronicles:
Tales of a New Mexico Harvest

By Carmella Padilla
Photography by Jack Parsons

Foreword by Stanley Crawford

Museum of New Mexico Press

Copyright © 1997 Museum of New Mexico Press. Text © 1997 Carmella Padilla; photographs © 1997 Jack Parsons.
All rights reserved. No part of this book may be reproduced in any form or by any means whatsoever,
with the exception of brief passages quoted for review, without the expressed written consent of the publisher.

Project Editor: Mary Wachs

Design and Production: Mary Sweitzer

Typography: Set in Janson with Bauer Bodoni display

Map by Beth Evans

Black-and-White Photo Conversion by Bonnie Bishop

Manufactured in Hong Kong
10 9 8 7 6 5 4 3 2 1

Library of Congress Cataloging-in-Publication Data Available.
ISBN 0-89013-313-1 (C); 0-89013-351-4 (P)

MUSEUM OF NEW MEXICO PRESS
Post Office Box 2087
Santa Fe, New Mexico 87504

Contents

Ristras of chile drying in Santa Cruz, New Mexico, ca. 1947–48. Museum of New Mexico Neg. No. 41279.

Foreword

Some years ago, I took to growing chile in the Embudo Valley to supplement our then basic crop, a local strain of top-setting garlic. Chile has always been a prominent feature in every traditional Hispanic garden, along with white corn and squash, and it was the major crop of several of the older Hispanic farmers from whom I had learned how to run an acequia in the early 1970s.

As a crop, chile also has claims to being indigenous, or relatively so. In climatic terms, the Embudo Valley between Española and Taos is about as far north as you can get, at six 6,000 feet, and expect to harvest a crop. What particularly interested me were the perennial discussions of the differences between Dixon chile, Chimayó chile, Española Valley chile, and Hatch chile by those who had more refined, or tougher, palates than my own. I was less interested in determining which was the better flavor, about which we all know there can be no arguing, than in the fact that climate and soil and breeding can give rise to remarkably distinct characteristics.

It seemed to me that here was matter around which to form a northern New Mexico chile growers' association to define and promote the various "land races," as they are called, that have evolved in the narrow river valleys of northern New Mexico since the conquest. Selling chile crops by year and chateau, you might say, or *año y huerta*, might seem at first like a too clever marketing idea, but if it encouraged the old farmers to pass on their heirloom varieties to the next generation and to keep their fields in agricultural production, then it could also be a lifeline for beleaguered small-scale agriculture in northern New Mexico.

In the pages that follow, Carmella Padilla covers chile production in the whole state of New Mexico while also seeing it clearly as one of the anchors of traditional Hispanic culture in these breezy times. Northern New Mexico may be one of the few places left in the country where a quasi-indigenous agriculture is still practiced, one that employs irrigation methods, crops, and cultural practices that predate, in part, the nation's various European colonial periods, an agriculture that is consummated by a revivifying cuisine developed by a people with ties of language and race to precolonial Mexico and the Pueblo tribes of New Mexico. Few other areas of the country have so successfully resisted or deflected or transformed the cultural flood tide of Anglo-European culture. The state forms the center of a culinary bioregion, with chile at its flaming heart.

My chile-growing years, perhaps five in all, were exciting but painful. I loved working with the plant and observing the startling transformation of the long green pods into wrinkled bright red husks, with intermediate blushes of purple and lavender and deep red. But harvesting the low-growing plant, which needs to be visited again and again, was brutal—for this particular six-footer. I never quite figured out how to stand on my head comfortably and look up from below into the canopy of drooping green leaves in search of the ripening pod to snap off. Some skills are best learned young from an elder so practiced that they seem natural and easy. Chile picking may be one of those. And the beauty of the ripening chile pods offers no resistance to the frequent early frosts in the far north.

Carmella Padilla has mapped out one of the most vibrant strains of Hispanic culture in northern New Mexico and traced how it has grown so wide and strong in the southern part of the state that it has influenced tastes throughout the nation and indeed the entire world. Chile is synonymous with New Mexico, north and south, yet the chile from northern New Mexico grown in small family *huertas* can seem to some like an entirely different crop from the commercial strains developed by the famous chile breeders of New Mexico State University in Las Cruces.

Some crops lure a person, or a people, into projecting their desires and hopes upon them, and gradually they yield to the human imperative to become what is desired. Difficult, stubborn, attractive, fiery, chile is one of those, perhaps the best of all. I still regret not having learned how to stand on my head for its sake, and each spring I think that maybe I ought to try again.

Stanley Crawford, El Bosque Garlic Farm,
Dixon, New Mexico

Acknowledgments

The people who participated in or otherwise influenced this book are numerous, and we gratefully acknowledge them here.

First and foremost, our thanks to the chile growers and their families who, during the course of four years, allowed us onto their farms and into their lives. Without their patience, hospitality, and enthusiasm, this project could have never been possible. Among them are: Jody Apple of Chimayó; Elizabeth Berry of Abiquiú; Eremita and Margaret Campos and Celia Leyba of Embudo; Orlando Casados, Sr., and Orlando Casados, Jr., of El Guique; Emma Jean, Dino, and Kristina Cervantes of Vado; Cordelia Coronado and Doña Agueda Martinez of Medanales; Robert and Barbara Cosimati of Las Cruces; Eliseo and Margaret Flores of Hatch; Rudy Herrera of Nambé; Faron, Jim, and Jo Lytle of Hatch; Gonzalo and Ermenda Martinez of Chimayó; Richard ("Dickie") Ogaz of Derry; Gary Riggs of Garfield; June Rutherford of Salem; and Amadeo and Rogie Trujillo of Nambé Pueblo.

Dr. Paul Bosland of the New Mexico State University Department of Agronomy and Horticulture was an invaluable resource who graciously gave us access to all aspects of the school's Chile Pepper Breeding Program and offered his specialized scientific insight into the wonders of chile.

Charles Gore, state statistician with the New Mexico Agricultural Statistics Service, patiently provided a thorough statistical picture of chile production in New Mexico.

Jeanne Croft, marketing specialist with the New Mexico Department of Agriculture, and Javier Vargas, county extension agent with the Doña Ana County Agricultural Extension Service, suggested valuable contacts in areas of chile research and growing.

Hatch residents Butch Anderson; Robert Duran; Johnnye Hammett; Jesús, Esperanza, and Yanet Perez; and Veronica Valdez gave us a unique glimpse of their famous chile community.

Donna Pierce and Robert Mayer gave tremendous scholarly and moral support to the author throughout the course of the project.

The Museum of New Mexico Photo Archives and the Rio Grande Historical Collections of the New Mexico State University Library contributed an important historical element to the book visuals.

Finally, special thanks to Museum of New Mexico Press Editorial Director Mary Wachs and to designer Mary Sweitzer and author Stanley Crawford. With Wachs's editorial expertise, Sweitzer's graceful design aesthetic, and Crawford's distinctive literary voice, this book has truly become everything we wanted it to be.

Among the books the author relied on for research during the course of the project are: *Peppers: The Domesticated Capsicums*, by Jean Andrews (University of Texas Press, 1984); *The Great Chile Book*, by Mark Miller with John Harrison (Ten Speed Press, 1991); *Chilies to Chocolate: Food the Americas Gave the World*, by Nelson Foster and Linda S. Cordell, eds. (University of Arizona Press, 1992); *Peppers: A Story of Hot Pursuits*, by Amal Naj (Knopf, 1992); *Sabino's Map: Life in Chimayó's Old Plaza*, by Don Usner (Museum of New Mexico Press, 1995); *Hispanic Villages of Northern New Mexico*, by Marta Weigle (The Lightning Tree, 1975); *Of Chiles, Cacti and Fighting Cocks: Notes on the American West*, by Frederick Turner (North Point Press, 1990); *New Mexico: A History*, by Marc Simmons (W.W. Norton, 1977); and *The Place Names of New Mexico*, by Robert Julyan (University of New Mexico Press, 1996).

The 1996 *New Mexico Chile Production Manual*, *The Chile Institute Newsletter*, and numerous other research articles provided by New Mexico State University's Cooperative Extension Service, Department of Agronomy and Horticulture, and The Chile Institute all served as invaluable resources as well.

Ristras of chile drying in Rio Arriba County, ca. 1920. Museum of New Mexico Neg. No. 31507.

Introduction

"Weeds," my father said, shaking his head. Spoon in hand, he paused before going in for another scoop of the steaming green chile stew. "*Malas yierbas.*"

I am a young girl, seated at the dinner table, surrounded by family. Outside the large picture window that lets the world into our Santa Fe kitchen, snow is falling in diagonal sweeps through the dark December sky. Inside, the heat from my mother's cooking clings to the cold window surface, causing droplets of water to form and roll down the glass. A warm veil of steam hangs in the kitchen air alive with the aroma of the dishes my mother has prepared. The slightly sweet smell of pinto beans unfolds upon the smoky scent of green chile and pork. I place a cold slice of butter at the center of a warm tortilla and watch it melt.

Manuel and Lorenzita Lujan, the author's grandparents, with chile from their 1952 harvest.

There is a feeling of comfort in my mother's kitchen, in the privilege of having a hot meal on a cold day. In my family, however, the true test of a hot meal is whether or not the chile is pungent enough for my father's fiery tastes. Dad is a voracious chile eater—as the extra freezer full of green chile, the dried chile *ristras* hanging in the garage, and the permanent red chile stains on various articles of his clothing attest to. My mother, on the other hand, doesn't care that much for chile, especially if it's hot. Yet she meticulously peels an entire bag of green chile or cleans a pile of dried red pods and cooks them into the family chile.

One of my earliest memories is of that December dinner, watching my father bite into a batch of chile that, for reasons known to God and nature alone, didn't bite back. Mom's stew tasted delicious as always, and my father ate every last drop. But on the heat scale, he may just as well have bitten into weeds. *Malas yierbas*.

Another memory brings forth the image of my father sitting at the same dinner table, eating a bowl of chile with my late Uncle Bill. Bill was as Texan as they come, but his idea of good chile was much more akin to my father's hot New Mexican version than his native Texas slurry of spicy meat and kidney beans. I loved to watch them sitting there together, sweat running down their brows and tears welling in their eyes, eating the hottest chile they could take. "Damn, that's good stuff!" one or the other would say, patting a handkerchief over his dripping face and neck. Then, no matter how painful to the palate, both would take another bite.

And finally, there was Thanksgiving, a day when my family's turkey and mashed potatoes were as likely to be smothered in red chile as brown gravy.

Such scenarios say something profound about the New Mexican culture into which I was born. Chile has played a central role in that culture for at least four hundred years, from the Spanish conquest of the region to the territorial period to statehood to today. During that time, it evolved from a unique culinary taste to a unique cultural tradition. Just as tea holds a place of prestige in Japanese culture and ritual, chile takes an integral, and often ceremonial, place in New Mexico life. Men, women, and children all share in the ritual, eating chile from morning to night, on ordinary days and on holidays, in weather hot and cold. They engage in endless debates about how chile should be cooked, how hot it should be, and what part of the state it should come from. Some eat chile as medicine, proclaiming it the only cure for the common cold. Others crave it as if it were a real drug and, as with any addiction, they suffer withdrawal when they go without. For old-timers and newcomers alike, chile is an essential part of the initiation into New Mexican culture.

As a writer, it is the human aspect of any story that most inspires me to put words on paper. As a New Mexican born into a family of self-proclaimed chile connoisseurs, it is a spicy chile pod that most inspires me to eat. Those two impulses merged to motivate my own exploration of New Mexican chile. In the spring of 1993, I teamed up with photographer Jack Parsons to undertake a collaborative documentary work that examined, in words and photographs, New Mexico's unique chile culture. Over the next four chile seasons we set out along 350 miles of the Rio Grande, from the hilly terrain of northern New Mexico to the lush, low-elevation farmlands of the south. From Embudo to Abiquiú, Hatch to Las Cruces and beyond, we visited with farmers in their chile fields, some as small as a few rows, others as large as five hundred acres. We found these New Mexicans to be true salt-of-the-earth people, men and women as deeply rooted in the landscape as the crops they care for.

Chile is one of the few agricultural crops cultivated by northern and southern New Mexicans alike, and the state's chile growers are responsible for producing 60 percent of all the chile grown in the United States. From early March to late November, we watched as these farmers, each in their own way, nurtured thousands of singular chile seeds through the delicate cycle of life. In virtually every chile conversation we had, there was talk of tradition, of the importance of preserving the distinctive cultural and culinary qualities of the crop. But there was also talk of the changes that have occurred as New Mexico chile has evolved from its backyard beginnings to a nearly $300-million-a-year state industry. Today's chile culture is no longer just about growing a favorite food. As farmers and others close to the crop explain it, it is also about science, economics, and enterprise.

At New Mexico State University in Las Cruces, scientists are busy breeding new varieties of chile to ensure the growth of specific colors, sizes, and shapes that appeal to the mainstream marketplace. These days, a chile ristra is as much a fashion statement as a food staple, and thanks to modern chile science, purple, yellow, and orange "ornamental" chiles are now as common as the standard red and green. The greater variety also has opened up a new commercial outlet for some farmers who specialize in growing chile not for taste but for color. The natural dyes the plant produces are extracted and sold to major manufacturers to color cosmetics, lunch meats, pet foods, and more. Scientists also have found evidence to explain why generations of New Mexicans have had faith in chile's natural healing powers. It turns out that a single chile pod contains up to six times as much vitamin C as a juicy Valencia orange; that's 130 percent of the U.S. Food and Drug Administration's recommended daily allowance in one medium-sized pod.

As little as fifty years ago, farming was a necessary means of survival in New Mexico. Today, many young New Mexicans are abandoning their family farming traditions to pursue college educations, better-paying jobs, and urban life-styles. The ever-increasing costs of farming equipment and labor, coupled with stiff competition among growers, have left some small farmers struggling to survive in the shadow of large agribusiness.

The Chile Chronicles: Tales of a New Mexico Harvest is our attempt to document the trials and triumphs of those farmers who persevere in a way of life that has endured in New Mexico for at least four centuries. From countless stories, we chose more than a dozen farmers and farm families to represent histories about and approaches to growing chile. Together, their various voices and landscapes comprise a portrait of New Mexico's most famous food.

For me personally, this book is also a picture out of my own past. In a faded photograph dated October 20, 1952, a man in a worn work hat and a woman in a dotted dress stand surrounded by ristras of red chile at least as tall as they are. It is a picture of my maternal grandparents, Manuel and Lorenzita Lujan, teachers who, on the side, farmed chile on a 160-acre ranch in Buford, New Mexico. It wasn't until long after my grandparents had died and I was well into this project that I discovered the photograph. As I began to learn about my own family's chile tradition I began to truly understand this land where farmers honor that tradition seed by seed, season after season, year after year. It is a place where there is always a chile tale to tell.

For my parents, Reyes and Zenaida,

who taught me to treasure the taste and the tradition of chile

and encouraged my dream of writing about it.

All my love, Carmella

For Becky, Chris, and Alex,

and for Mac and Marjorie,

with love, Jack

Chile ristras drying in northern New Mexico, ca. 1935. Photograph by T. Harmon Parkhurst. Museum of New Mexico Neg. No. 155349.

[Chapter One]

A History of New Mexico Chile

In the beginning was the chiltepin.

Genesis, in the book of New Mexican chile, takes place at least nine thousand years ago in South America, among the indigenous peoples of Bolivia and Peru. There, tiny, pungent red berries, no bigger than the rounded head of a hatpin, grew wild upon low-lying vines that stretched across the jungle floor. In Peru, the berries were among the principal foods of the Incas, who referred to them as *Uchu*, the name of a prominent character in one version of the Incan creation myth. In Central America, the Mayans swallowed the egg-shaped berries to treat stomachaches, while in Mexico the Aztecs called the fiery fruit "chilte*c*pin," or "flea chile" in their native Nahuatl language. So highly regarded were these peppers that whenever they wished to mollify their gods by fasting, the Aztecs sacrificed two things: sex and chiltepin.

The spread of the chiltepin north from South America to Central and North America owed as much to the migration of birds as of people. Birds are naturally attracted to red fruit and, unlike most other small mammals, are equipped with digestive tracts that chemically soften seed coats without damaging the potential for germination. Humans, too, traveled the region, though they weren't quite so generous with their wild chile seed. They exchanged them with foreign traders in return for other goods.

The sequence of events that comprises the history of chile is as scattered and inexact as the seeds that pre-Columbian birds deposited upon the Americas. Old and new theories abound about the origins of chile, a fruit of the genus *Capsicum*, but in recent years, scholars, botanists, horticulturists, and other researchers have begun to piece together the evolution of this important food.

Chile ristras drying, Chamita, New Mexico. Museum of New Mexico Neg. No. 5181.

Whether or not the Incas truly believed that their civilization was spawned from chile is unclear. However, archaeological evidence shows that South Americans were growing chile as a food between 5200 B.C. and 3400 B.C. and that in Mexico it was being eaten as early as 7000 B.C. Before 1492, when Christopher Columbus set sail for the New World in search of spice, chile wasn't known in Europe. Columbus was looking for black pepper, at the time as valuable a commodity in the European marketplace as silver. What he found instead was the chiltepin. The plant's small red berries resembled black pepper in its ripened stage, and it possessed a spicy heat similar to *Piper nigrum*, the source from which pepper is derived. Columbus was confused; just as he erroneously called the New World natives "Indians," he mistakenly called the plant "pepper." His poor choice of words would complicate the lives of botanists for generations to come.

Nonetheless, chile seeds were among the bounty Columbus carried back to Iberia, where they assumed a prestigious place in the local cuisine alongside pepper itself. During the next hundred years, Spanish and Portuguese explorers took the seeds along on expeditions to countries as diverse as Africa, India, China, Hungary, and Tibet. In every place, Columbus's distinctive New World pepper was praised as a substitute for the popular Old World spice. Chile soon became so thoroughly incorporated into the cuisines of India and Indochina that many early botanists believed the plants originated there, although scholars eventually would designate South America as the plant's ancestral home.

It was Mexico, though, that would get the credit for domesticating the various types of chile that evolved from the chiltepin into the species *Capsicum annuum*. By the time Spanish explorers, led by Hernán Cortés, began their invasion of Mexico in 1519, Aztec plant breeders already had developed dozens of cultivars within the species, including poblanos, jalapeños, and serranos, as well as other nonpungent chile plants. In its native South American home, the word for chile was *aji*. The Aztecs, however, used the Nahuatl term *chilli* to refer to various chile cultivars. Perhaps the most sophisticated chefs of their time, the Aztecs laid the foundation for modern Mexican and New Mexican foods by using chile in the preparation of moles, tamales, salsas, pipián, and other dishes and sauces. A 1569 history by Spanish friar Bernardino de Sahagún documenting the daily life of the Aztecs before the Spanish conquest lists among the diverse foods: "Frog with green chillis; newt with yellow chilli; tadpoles with small chillis; maguey grubs with a sauce of small chillis; lobster with red chilli" (in *Peppers: The Domesticated Capsicums*, by Jean Andrews). Upon their arrival, the Spaniards called the plant *chile* in their own language.

In 1540, the Spanish conquest reached New Mexico in the expedition of Francisco Vásquez de Coronado, and within the century chile seeds were given to the indigenous Pueblo Indians. By 1598, shortly after the Don Juan de Oñate expedition reached northern New Mexico, chile was being grown using water from the Rio Chama and advanced methods of irrigation introduced by the Spanish. In its native tropical home, chile had grown as a perennial shrub and could live for a decade or more, but in New Mexico a killing frost destroyed the plants each fall. The Spanish thus cultivated the crop as an annual, reestablishing the plants every year and saving a portion of the seeds for the next year's crop.

Collecting dried red chile in a field at the Barker farm near Las Cruces, ca. 1950. Photograph by Maurice Eby. Rio Grande Historical Collections, New Mexico State University Library.

New Mexico's high-desert climate—warm days, cool nights, and steady breezes—proved perfect for growing chile. The Spanish method of irrigation employed a series of *acequias*, or ditches, that directed water to the chile fields, helping to supplement the little rainfall in the region. Such methods became key to the crop's success. The Pueblo Indians also planted chile alongside their traditional crops of beans, squash, and corn, and they combined the foods in inventive new ways. They ate chile green and red, fresh and dried, hot and mild. They even ate it to cure colds. Before long, chile had become an essential ingredient in the lives of both the Spanish and Indian cultures of colonial New Mexico.

Chile grew along the Rio Grande in northern and southern New Mexico for at least two centuries before most of North America had even heard of the crop. With the opening of the Santa Fe Trail in 1821, chile was introduced to other parts of the United States. With its export came the anglicized name "chili," retained outside the Southwest today. The arrival of the railroad to the region in 1880 brought even more Americans to the chile table.

The Chili Line, a narrow-gauge branch of the Denver and Rio Grande (D&RG) Western Railroad, was constructed in early 1881 as part of William Jackson Palmer's plan to extend his Colorado railroad empire

The Chili Line brought chile and other products to markets throughout the West. It blew its last whistle on September 1, 1941. Photograph by Margaret McKittrick. Museum of New Mexico Neg. No. 41833.

south from Denver to Mexico via Santa Fe. By the end of that year, the new branch reached as far as Española, but an earlier agreement signed by the railroad with the Atchison, Topeka, and Santa Fe Railroad Company kept the line out of the Santa Fe area. Chili Line passengers were forced to board the bumpy Barlow and Sanderson stagecoach for the final thirty-four miles from Española to Santa Fe. However, a group of Santa Fe merchants desperate to link their city to marketplaces in the North soon took matters into their own hands. They founded the Texas, Santa Fe, and Northern Railroad Company, which, by 1887, completed the remaining forty miles of rail to Santa Fe. The squat Mikado steam train became a favorite of tourists on excursions from Santa Fe to nearby Indian pueblos. Traveling through the timeless spaces of northern New Mexico was to travel in a foreign land. Conductors collected fares in both English and Spanish. Signs on the bathroom doors said "*Mujeres*" and "*Hombres,*" with "Women" and "Men" printed like afterthoughts in small letters below.

It was a comfortable ride. The train cars were lined with oak paneling, their interiors brightened by the warmth of potbellied stoves and the light of brass-and-crystal kerosene lamps that swung from above. Besides the tourists, passengers included a constant flow of sheepherders who traveled to and from their seasonal jobs in Wyoming. During World War I, thousands of soldiers were shipped out of the region on the line. The Chili Line moved timber to markets throughout the West and carried minerals, such as gold, copper, and quartz, and livestock as its freight. Every fall, the train was filled with the agricultural bounty of northern New Mexico. Along with carloads of apples and piñon nuts, strings of chile were transported north. The brilliant red ristras that were hung to dry upon the adobe facades of the villages that lined the train's course inspired the Chili Line's name.

Fresh green chile roasting in an* horno *in Jarales, New Mexico, 1939. Photograph by Irving Rusinow. Museum of New Mexico Neg. No. 9367.

The most common variety of chile grown in the United States today is the bell pepper, but in New Mexico several hundred other varieties are cultivated. The three major chile types grown in the state are New Mexican, jalapeño, and cayenne. These pod types are further divided into three categories: green chile, red chile, and paprika.

The train occupied a prominent place in the isolated lives of the people of the villages through which it passed, places with lonely names such as *No Agua* ("No Water") and *Tres Piedras* ("Three Rocks"). Every day at the designated times, villagers waited alongside the tracks as the engineer slowed the train and a sack of mail, a newspaper, or some other news of the faraway world sailed through the air and landed at their feet, bringing a brief glimpse of life beyond the confines of a still-isolated northern New Mexico. But for all its beauty and romance, the train's route was also treacherous, earning the D&RG branch a nickname as the "Dangerous and Rough-Going." Spanning some 125.31 miles between Antonito, Colorado, and Santa Fe, the train traversed great lengths of rocky, rugged terrain highlighted by a series of winding hills, curves, dips, and loops. Just north of Embudo, Barranca Hill propelled the train down a seven-and-a-half-mile drop of 1,128 feet. It took forty-six minutes and nothing short of a daily engineering miracle to ensure that the train descended without careening off the tracks.

In 1939, when the D&RG Western Railroad announced its plans to abandon the Chili Line, a series of protests ensued. The railroad company, which had been bankrupt and in the hands of a receiver since 1935, claimed that passenger traffic had dwindled to a mere ten passengers per trip. But the New Mexico and Colorado businesspeople and villagers who depended on the train argued that halting service would strike a severe blow to the economy of the region. The debate became so heated that a special congressional subcommittee was convened to determine the significance of the Chili Line. The subcommittee recommended the line be maintained for purposes of national defense, but the Interstate Commerce Commission wasn't convinced and, on August 21, 1941, authorized abandonment of the Chili Line. On September 1 of that year, Engine No. 470 puffed out of Santa Fe for the last time with twenty passengers on board, leaving the capital city without regular rail connections for the first time in sixty years. With the death of the Chili Line, twenty-four engineers, conductors, brakemen, and firemen lost their jobs. Hundreds of farmers and merchants lost their way into the national marketplace, and New Mexico lost the first, and perhaps the greatest, public relations tool for chile the state would ever have.

Storing dried red chile at the Barker farm near Las Cruces, ca. 1950. Photograph by Maurice Eby. Rio Grande Historical Collections, New Mexico State University Library.

In 1862, President Abraham Lincoln signed the Morrill Act directing legislatures throughout the West to designate a university in every state or territory as an official land-grant college. America at the time was still a nation of farmers, and such a mission was intended to provide agricultural research and extension services on their behalf. Thus was established in 1889 the New Mexico College of Agriculture and Mechanic Arts in Las Cruces. Fabian Garcia was one of five students in the school's first graduating class of 1894.

Garcia, a native of Mexico, had a keen interest in the principal crops grown in the area, including cotton, onions, and pecans. But the one that most caught his attention was native to his own Mexican home: chile. After graduation, Garcia was hired as a researcher at the school's Agricultural Experiment Station. There he began to compile the scientific research that would lead to the development of a new chile unique to New Mexico.

Garcia based his initial research on a feeling. Although chile was a major crop in New Mexico, it was still largely being grown—and eaten—by Hispanics. At the time, chile also was being canned green in California, used in Texas to spice up a bowl of meat and beans, and turned into hot sauce in Louisiana, but the New Mexican practice of eating chile as a main dish was virtually unknown. Garcia's feeling was that chile could become a major commercial industry if it was adapted to more mainstream American tastes. He believed that New Mexico needed a new chile cultivar more uniform in size and milder in taste than the standard Mexican chile then being grown in the area. If he could create such a chile, Garcia was sure he could hook the territory's growing Anglo population, and eventually the rest of America, on the New Mexican crop.

In 1908, Garcia published an article entitled "Chile Culture," in which he discussed his ambitious plans to create a contemporary New Mexican chile. One year earlier, he had begun experiments to cross the *Chile Pasilla* plant with the *Chile Colorado* plant. Also known as the *Chile Negro* and literally translated as "little raisin," the dark, chocolate-colored Chile Pasilla pod was long, tapered, and as wrinkly as a raisin. The Chile Colorado pod was about the same size and shape as the Pasilla but, when mature, its skin turned a brilliant shade of scarlet.

By 1912, the year New Mexico became the forty-seventh state in the Union, Garcia's annual experiments had not yet resulted in a New Mexican pod type. But that same year, pharmacist Wilbur L. Scoville invented the Scoville Organoleptic Test to measure the amount of heat in a chile pod. The degree of heat, or pungency, in chile is determined by the amount of capsaicinoids, a mixture of seven closely related compounds, found in the fruit. The capsaicinoids are concentrated in the placenta tissue, or vein, of a chile pod and can be detected by human taste buds in solutions of one part per million. Scoville relied on trained testers to provide a subjective measure of chile heat; for every one part per million concentration of capsaicinoids they detected on a particular pod, 15 Scoville units were assigned. The measurements ranged from 0 Scoville units for a bell pepper, a nonpungent chile type that contains no capsaicinoids, to up to 200,000 Scoville units for the blazing habanero, grown in the Yucatán and considered one of the hottest chiles in the world.

Finally in 1921, fourteen years after planting his first crossbred chile, Garcia introduced "New Mexico No. 9" as the first scientifically developed chile cultivar. (The rationale for the chile numbering goes not to relative heat but to the vagaries of growing.) The new pod type was longer, milder, and redder than the common local chile, and, much to the delight of area farmers, its seeds produced high yields. New Mexico No. 9 would be the standard cultivar grown in the state for the next thirty years.

Meanwhile, in northern New Mexico, the descendants of the first chile seeds planted in the region by the Spanish in 1598 had flourished into distinctive cultivars of their own. Identified by horticulturists as "land races," these seeds had evolved naturally into specialized varieties of chile grown only in certain geographic areas. Some of the most prominent land races developed in places such as Dixon, Velarde, and Chimayó and thus were named for these areas. Unlike the chile that Fabian Garcia had scientifically designed for uniformity, these chiles were small, crinkly, and far from uniform, a reflection of the irregular climate, soil, and geography of their northern home. Their thin skins made them harder to peel than Garcia's tailor-made chiles, but they tasted distinctly different and, according to many, distinctly better than the southern pods.

By 1937, a study on the production and marketing of chile in northern New Mexico by the U.S. Department of Agriculture estimated that some sixty thousand ristras were produced annually in northern New Mexico villages for trade at area mercantiles. The study concluded by saying: "There clusters around chili not merely a set of agricultural practices, but a total institution. Any work undertaken . . . in the future that involves chili will invite failure if it is not recognized that the ramifications of chili are extensive—they extend not only to the native economy but through the economy to the total culture."

By the time he retired in 1946, Fabian Garcia had authored more than twenty articles on the cultivation of chile and had laid the groundwork for turning a regional food into a national food. Above all, he had made chile into a science. His legacy inspired other researchers to continue to develop new and improved New Mexican chile types more palatable to American tastes and more profitable for New Mexico farmers.

Among these researchers was Roy Harper, who in 1950 unveiled "New Mexico No. 6" as an even milder—50 to 60 percent milder—alternative to Garcia's New Mexico No. 9. New Mexico No. 6 produced higher yields than the Garcia cultivar, and its smooth, thick-fleshed pods were well-suited for canning. In 1956, with the release of his "Sandia," a hot, intermediate-sized pod, Harper proved that he could work at the other end of the

Fabian Garcia, the father of the New Mexico chile industry. He was responsible for the first scientifically developed New Mexican chile. Rio Grande Historical Collections, New Mexico State University Library.

heat spectrum, too. Still, it was Harper's early testing of a pod type known as "New Mexico 6-4," a descendant of his New Mexico No. 6, that would prove most popular among chile growers. Harper began testing the medium-hot pod type in the early 1950s in hopes of developing a chile whose quality would endure from the early green through the mature red stage.

By the time New Mexico 6-4 was released in 1958, Harper had been succeeded at the college by a researcher named Roy Nakayama, the son of Japanese farmers who settled in the community of Doña Ana in southern New Mexico around 1918. Nakayama, like Harper, was a researcher at the Agricultural Experiment Station where he perfected Harper's New Mexico 6-4. Measuring six to seven inches long with a smooth skin and thick flesh, the new pod type was immediately attractive to growers because of its adaptability for both processing and dehydrating. As the regional commercial chile market continued to expand nationally, New Mexico 6-4 would become the most widely grown chile in the state.

Approximately 1,200 acres of chile were grown annually in New Mexico between 1949 and 1959. In 1965, during the first session of the Twenty-Seventh New Mexico State Legislature, lawmakers passed Bill No. 24. It declared the chile, along with its most common culinary partner, the pinto bean, the official vegetable of the state. Two years later, Nakayama introduced the large, medium-hot "Rio Grande 21" as the latest genetically engineered chile to come out of the New Mexico College of Agriculture and Mechanic Arts. Although the college's agricultural mission remained strong, legislators in 1969 adopted a state constitutional amendment that changed its name to New Mexico State University (NMSU). Some 7,780 acres of chile were planted in the state in 1973. By 1975, that figure had jumped to 9,200 acres. The New Mexico chile boom had begun.

NMSU's chile breeding program was thrust into the national spotlight with Nakayama's 1975 release of the "NuMex Big Jim" chile. The king-size chile also earned June Lytle Rutherford, the Hatch farmer who grew it for Nakayama, a spot in the *Guinness Book of World Records* for growing the biggest chile—13.5 inches long—ever recorded. Nakayama had both the backyard gardener and the chile cook in mind when he crossed pollen from a tiny Peruvian chile with other New Mexican varieties to develop the Big Jim. The large, medium-hot chile was easy to grow, easy to harvest, and, most important, easy to stuff when making *chiles rellenos*.

By 1979, Nakayama was known internationally as "Mr. Chile," and the state's chile production had increased to nearly fifteen thousand acres. Four years later in Washington, D.C., New Mexico's Sen. Pete Domenici took the floor of the U.S. Senate in an effort to educate his fellow lawmakers about "the correct way to spell chile." The "e" ending in chile is the authentic Hispanic spelling of the word, the senator explained, whereas chili with an "i" serves only to identify the official state dish of Texas. That said, the senator returned to his seat knowing he had served his constituency well.

Roy Nakayama's tenure with the university spanned nearly thirty years until his retirement in 1985. He spent most of his final years at NMSU trying to develop a chile whose pods would grow upward in clusters so they could be easily machine-harvested. Nakayama never accomplished that goal, but he achieved others. For the farmers in northern New Mexico, where the growing season is too short for slow-maturing chile, he created his 1984 "Española Improved," a hot, fast-maturing chile that was cultivated alongside the established land races of the region. Then, just before his retirement, he unveiled the 1985 "NuMex R Naky," the first scientifically developed, low-pungency chile. This particular pod proved the perfect source for paprika, a very mild and very red chile powder, which at the time was just gaining ground as a natural coloring agent. Before Nakayama's paprika, the United States imported the product from such places as Hungary and Spain. With the release of R Naky, a new commercial market opened up for New Mexico chile growers, who soon exported an estimated one-third of their homegrown product to Hungary, Spain, Africa, and other markets abroad.

From Garcia to Harper to Nakayama, chile research at New Mexico State University had shifted from efforts to get the best green and red chile out of a single plant to a focused study on how to grow both green and red chile for singular use. In 1986, when Paul Bosland replaced Nakayama as the university's chile breeder, the focus shifted again, this time toward the development of multicolored "ornamental" chiles that would serve the growing market for ristras, wreaths, and other chile decor. Among Bosland's colorful creations are the yellow "NuMex Sunrise," the orange "NuMex Sunset," and the purplish brown "NuMex Eclipse," all released in 1988. In 1989, NMSU marked its one-hundred-year anniversary, and the state chile crop expanded again, to 23,500 acres. In honor of the occasion, Bosland introduced the "NuMex Centennial," a purple potted piquin pepper geared for use in the greenhouse and floral industries.

In the tradition of his predecessors, Bosland's research also centers on the creation of disease-resistant chiles, as well as on preservation techniques

Green and red chile represent two developmental stages of the same fruit. The flavor of green chile is completely different than that of red chile because the pods are picked at different physiological stages. When green, chile is a vegetable. When red and dried, it is a spice. But botanically, chile is a berry. It is a member of the Solanaceae, or nightshade, family, a kin to plants as diverse as tomato, potato, tobacco, and petunia.

From the mid-1950s to 1985, researcher Roy Nakayama developed "NuMex Big Jim," "Española Improved," and other chiles to meet a growing culinary demand. Rio Grande Historical Collections, New Mexico State University Library.

During a special session of the Forty-Second New Mexico State Legislature in Santa Fe, state legislators registered a vote of confidence and support for the New Mexico chile industry when House Joint Memorial 3 breezed through both the state House of Representatives and the Senate. The memorial recommended that the question "Red or green?" be proclaimed the official question of New Mexico. The phrase refers to the question asked in New Mexico homes and restaurants countless times every day: "Would you like your chile red or green?" On April 10, 1996, Secretary of State Stephanie Gonzales issued a proclamation declaring "Red or green?" the quintessential state query.

intended to give green chile an edge in the fresh produce market. In 1990, Bosland moved one step closer to creating the perfect green chile with the "NuMex Joe E. Parker" pod. A mild cultivar, the Parker chile not only produces outstanding green chile yields in the field but its thicker fruit wall results in more green chile on the plate after peeling. For those wishing to leave the chile in the fields to ripen, it produces exceptional red chile yields as well. Today, the Parker pod has replaced Nakayama's New Mexico 6-4 as the most widely grown chile in New Mexico.

The ongoing innovations in chile at NMSU have earned the school an international reputation as one of the world's most progressive and prestigious institutions for chile research. The Las Cruces campus has developed more of the chile cultivars currently grown in the world than any other university, and advanced technology continues to help researchers devise new techniques in chile breeding and management, as well as brand-new uses for the plant. For instance, a High Performance Liquid Chromatograph now records an objective measure of chile heat. Yet to come is a chile that is harvestable by machine.

According to the U.S. Department of Agriculture, the amount of chile consumed by Americans each year almost doubled, from 3.5 pounds per person in 1980 to 6.5 pounds per person in 1993. In 1991, sales of salsa in the United States surpassed that of ketchup, which for decades reigned as America's favorite condiment. Former English professor and self-proclaimed "Chilehead" Dave DeWitt was living in Albuquerque in 1987 when the national chile craze inspired him to join forces with his Albuquerque friends Nancy Gerlach and Robert Spiegel in publishing *The Whole Chile Pepper Catalogue*, an everything-you-ever-wanted-to-know-about-chile regional publication. The successful catalogue soon gave birth to the bimonthly *Chile Pepper* magazine, devoted to the hot and spicy cuisines of New Mexico and the world. Before long, *Chile Pepper* reached a circulation of eighty thousand, prompting DeWitt to launch the annual Albuquerque Fiery Foods Show, a three-day affair bringing together thousands of hot-food producers from around the world.

By 1992, U.S. commercial chile acreage, some 125,000 acres, exceeded that of celery or honeydew melons while per capita use of chile surpassed asparagus, cauliflower, or green peas. The majority of that chile was being grown in New Mexico. The year 1992 was a record-breaker, with farmers harvesting a total of 34,500 acres. Straight off the farm, the crop was valued at $67.3 million; after processing, the value exceeded $250 million.

One year later, a long-silent rivalry was ignited between New Mexico and Texas, which claims not to grow the most chile in the country but the best. That year, New Mexico state legislators accepted a challenge from former Texas Governor Ann Richards to prove which state's chile is superior. The result was the start of an annual chile cookoff between chefs from the two states. During the first three years of the contest, New Mexico claimed the culinary crown. In 1996, Texas rebounded with wins in both the best chile and salsa categories.

Harvested chile acreage in New Mexico decreased to 27,900 acres in 1994, the lowest since 1989 and a 7 percent drop from the previous year. The decrease was largely attributed to an overproduction of chile in previous years due to growers overestimating its demand. By cutting back on their acreage, farmers attempted to bring the market back into balance. Even so, New Mexico remained the top producer of the crop in the United States, with Texas, California, and Arizona following in descending order.

An abundant chile and corn harvest in northern New Mexico. Museum of New Mexico Neg. No. 31509.

The year 1995 brought another drastic decrease in acres of harvested chile, this time to 22,400 acres and a total value of $44.8 million. With strong winds and the most destructive chile virus, the "Curly Top," making a widespread appearance in fields across the state that year, it was not human error but Mother Nature that was to blame. In 1996 yields were up and diseases down.

Today, nearly one hundred years after Fabian Garcia began his quest to make chile a major commercial industry in New Mexico, the state is responsible for producing 60 percent of the total chile acreage in the United States. Eighty percent of the state's chile is designated for the processing industry, where it will be frozen, canned, pickled, or dried. The other 20 percent will be sold through fresh market channels, such as grocery stores, farmers' markets, and roadside stands.

With the exception of the tabasco and the habanero chiles, nearly all chile grown in New Mexico today belongs to the same *Capsicum annuum* species that Spanish explorers discovered in Mexico almost five centuries ago. These same chiles are the most widely grown and most economically profitable the world over.

Native American–Spanish Chile Debate

Rudy Herrera, top, and Amadeo and Rogie Trujillo offer different perspectives on the origins of New Mexican chile. Opposite: A colorful sampling of chiles grown by Herrera.

Some forty years after Francisco Vásquez de Coronado's 1540 expedition in New Mexico, surveyors returned to record the findings. Chronicler Baltasar Obregon made an important observation: "They have no chile," he wrote, "but the natives were given some seed to plant."

Obregon was referring to the Pueblo Indians, who beginning with their Anasazi ancestors had inhabited the region at least since A.D. 860. Archaeological evidence has supported the early Spanish report, but the debate still rages between some Hispanic and Pueblo Indian locals as to the origins of New Mexico chile.

At Trujillo Family Farms in Nambe Pueblo, Amadeo Trujillo grows about one acre of chile each year in an effort to maintain what he believes is an indigenous crop of his Native American ancestors. Amadeo argues that by the time the Spanish arrived, the Pueblo Indians had been trading with Indians in Mexico, making it likely that chile was traded as well. And he balks at Spanish writings of sixteenth-century Pueblo life absent of chile, saying that even if the Spaniards had discovered chile here, they wouldn't have given the Indians credit for such a valuable crop.

"I think it all started at the pueblos," Amadeo says. "Maybe what the Spaniards brought were improved varieties. Then the chile spread to places like Chimayó, and everyone started thinking it was a Spanish crop. But chile was already here."

Meanwhile, in nearby Nambe village, not connected to the pueblo, Rudy Herrera plants a small chile plot every spring on his sprawling Rancho Poco Loco (Little Crazy Ranch) as part of his efforts to preserve the Hispanic way of life that has thrived in the Rio Grande Valley since his Spanish ancestors arrived. Rudy modeled his fifty-one-acre ranch after a traditional Spanish Colonial ranch, a virtually self-sustaining operation that functions as much as possible on the same principles of self-reliance that his forebears depended upon. Along with Christianity, Rudy believes the Spanish brought chile to the region, sowing the seeds that were the progenitors for the "native" chile grown at both Nambe Pueblo and in his village today.

"We can argue all we want about who had the chile first, but the fact is that chile is an important crop for all of us," Rudy says. "It not only is a main food ingredient, it's a main social ingredient. It literally keeps our communities together."

A field of chile at New Mexico State University is draped with insect-proof nylon netting, a technique that encourages plants to self-pollinate and helps chile breeders ensure genetic purity in seeds. Inset: New chile cultivars start from seedlings in trays and grow into healthy plants inside the New Mexico State University experimental greenhouse.

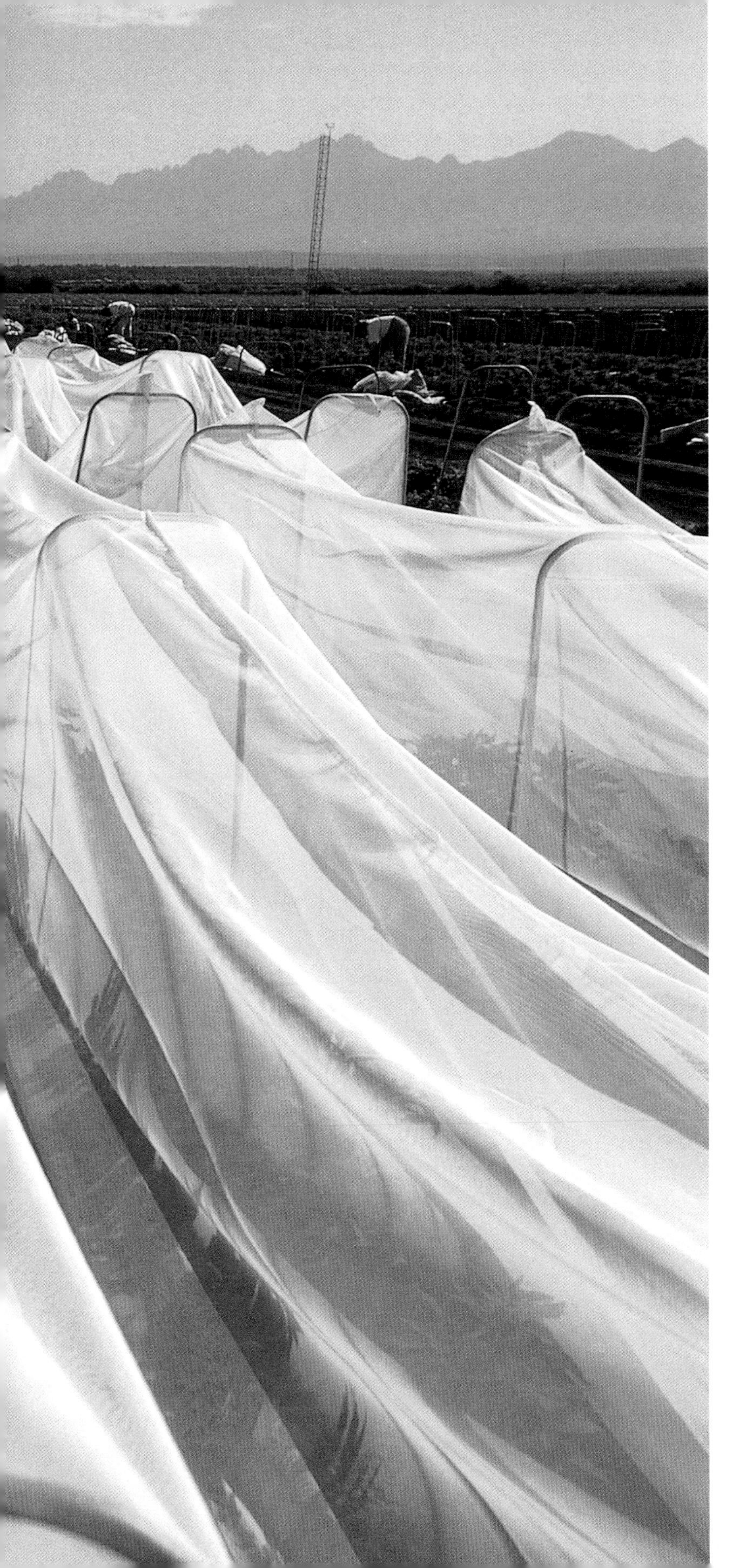

[Chapter Two]

Chile Science

The agricultural, culinary, and economic significance of chile in New Mexico today owes as much to the Chile Pepper Breeding Program founded by Fabian Garcia in the early 1900s at New Mexico State University in Las Cruces as to the farmers who produce the crop. Beginning with New Mexico No. 9, the original New Mexican chile created by Garcia in 1921, the university has been the birthplace for a long line of chile cultivars that has been adapted to meet the geographic and economic needs of chile growers, processors, and consumers. By 1996, nearly thirty of the newest chile cultivars grown in New Mexico and the world had been developed at NMSU, and ongoing research into the genetics, pathology, nutrition, color, pungency, marketability, cultural management, and other aspects of the chile plant has put the university at the forefront of chile research.

Vegetable breeder and horticulture professor Paul Bosland has headed New Mexico State University's prestigious Chile Pepper Breeding Program since 1986.

It takes seven years to breed a better chile. In that time, under optimal conditions, one can take the heat out of the most pungent pod, create a deeper shade of red, grow a thicker layer of meat beneath an easier-to-peel skin, or build resistance to the nasty fungi and insects that cause chile disease. The merging of desirable chile genes through classical crossbreeding techniques is now the dominion of Paul Bosland, NMSU horticulture professor, vegetable breeder, and current director of the program started by Garcia. The "Chile Doctor," as a sign in Bosland's office proclaims him, is the only remaining university-level chile researcher in the only chile breeding program of its kind in the country.

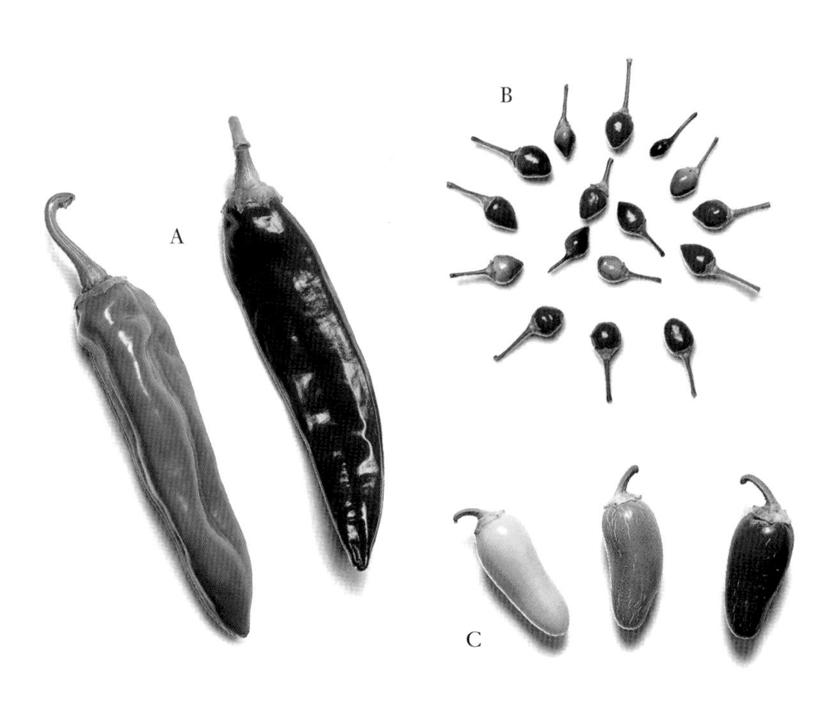

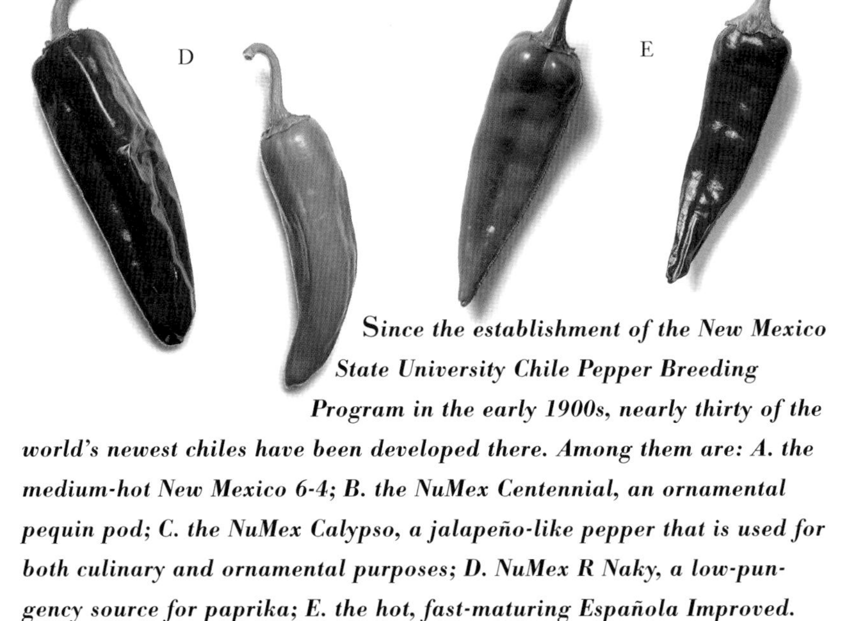

Since the establishment of the New Mexico State University Chile Pepper Breeding Program in the early 1900s, nearly thirty of the world's newest chiles have been developed there. Among them are: A. the medium-hot New Mexico 6-4; B. the NuMex Centennial, an ornamental pequin pod; C. the NuMex Calypso, a jalapeño-like pepper that is used for both culinary and ornamental purposes; D. NuMex R Naky, a low-pungency source for paprika; E. the hot, fast-maturing Española Improved.

As director of such a rare enterprise, Bosland has become a celebrity of sorts. Every year, he receives thousands of letters and phone calls requesting information about chile. The questions come from everyone, from backyard gardeners to fellow academics to writers to restaurant chefs. In the true spirit of scientific inquiry, Bosland seeks to answer every query, no matter how trivial or tough. A letter recently arrived from a farmer in Africa who, after learning that New Mexico's average annual rainfall is about equal to rainfall levels in his village, wanted information about how to start a chile crop. Bosland was sorry to reply that, given the absence of an irrigation system in the man's African home, it couldn't be done.

"Plant breeding is an art and a science," Bosland says while strolling through the university's one-acre experimental garden that each year hosts as many as one thousand different varieties of chile from around the globe. The pods, most of which belong to the *Capsicum annuum* species, the most widely grown chile pepper species in the world, wear glossy skins of purple, yellow, orange, white, chocolate, red, green, and black. In shape, the fruits range from the mushroomlike Scotch bonnet and the cherry-tomato Hungarian cherry pepper to the skinny long-stemmed tabasco, the fat waxy guero, the blocky bell pepper, and the beefy poblano. Grown erect or pendent, they go from tongue-biting to totally nonpungent.

"Tasting chile is like wine tasting," Bosland explains. The ancho, for instance, is sweetish, the mirasol fruity, the chipotle smoky, and the mulato hints of chocolate. A red Hungarian cherry pepper is sweet and mild with tomato undertones, while the brilliant orange habanero, similar to the Hungarian fruit in size, is blistering hot with a slightly acidic aroma. "When you first drink wine, all you taste is the alcohol. With chile, you taste the pungency. Eventually, you become a connoisseur."

Bosland's exotic selection of chiles provides valuable genetic material for making improvements in the New Mexican crop. Indeed, even the subtlest qualities of the worldly chiles grown in the agricultural garden are relevant to Bosland's breeding program, in which hundreds of new chile breeding lines are in progress at any one time. The program focuses on pushing the major New Mexico chile types—the New Mexican, jalapeño, and cayenne—to their fullest potential in terms of pungency, flavor, color, size, shape, and resistance to disease.

In the first half of the twentieth century, Fabian Garcia and Roy Harper sought to improve the New Mexican chile as a dual-purpose crop, one that produced both the best possible green and red chile out of a single plant. But beginning with Roy Nakayama in the 1960s, the breeding emphasis shifted to single-use crops. Today, some chiles are bred solely for

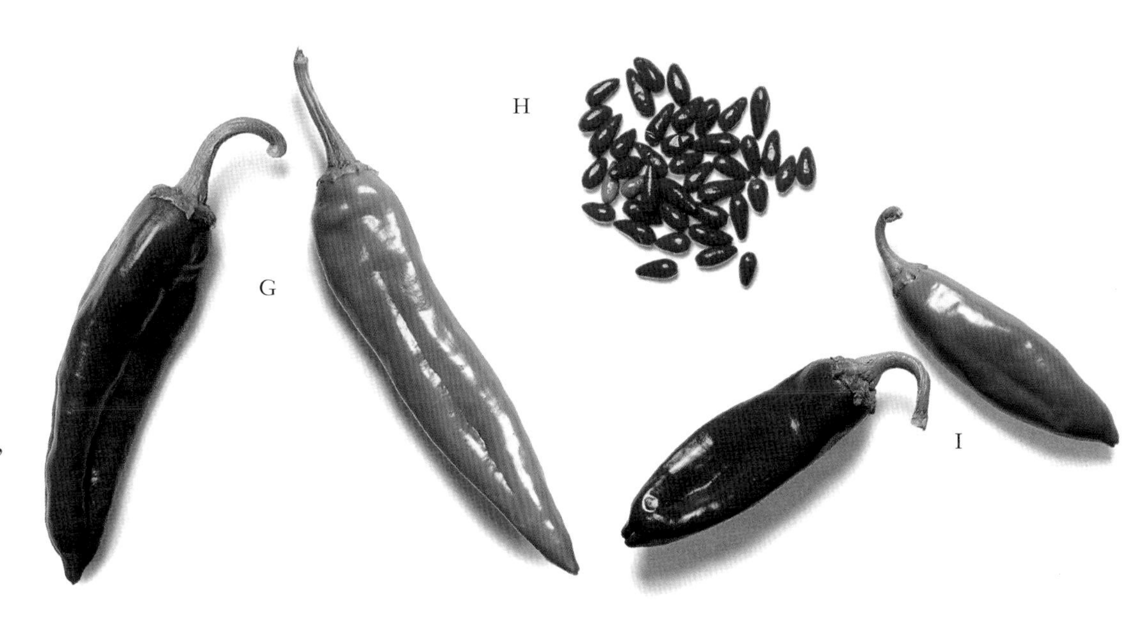

Chiles this page: F. the ornamental NuMex Sunburst (orange), NuMex Sunflare (red), and NuMex Sunglo (yellow); G. the king-size NuMex Big Jim; H. the NuMex Bailey Piquin, a hot berry-sized pod; I. the paprika-perfect NuMex Conquistador.

color or shape without any consideration given to flavor. Others are bred to satisfy very specific tastes. One can breed for a milder green, a hotter red, or an orange the color of a pumpkin. A chile can be made more meaty or more lean, with fewer or greater number of seeds. Ninety percent of research time, though, is devoted to developing resistance to the many diseases that destroy chile crops.

At 10 A.M. on a June morning, Paul Bosland arrives at a one-acre experimental chile field located about two miles from campus. Temperatures are close to 100 degrees F, yet a group of summer school students is hard at work in the soil among long rows of two-month-old chile plants. The rows are covered by long white swathes of nylon netting and are interrupted every few feet by wooden stakes, between which a different cultivar grows.

A key stage of the chile breeding process is under way. Pollination, or the process of sexual reproduction in plants, is the union of male and female reproductive cells to create seeds. Plants of the same species that originate from seed can be quite different from their parent plants or from one another. This fact enables chile breeders to make controlled crosses between chile plants, combining genes in new ways to create new cultivars.

From a breeding standpoint, the flowers that form on a chile plant are perfect; each has both male and female organs that are easy to distinguish and easy to cross. Pollination, the transfer of pollen from a plant's anther to a stigma, can occur two ways. In cross-pollination, the pollen is transferred from the anther of one plant to the stigma of another plant by insects such as bees. In self-pollination, the pollen is moved from the anther to the stigma of the same flower, or to the stigma of another flower on the same plant, by the plant itself. Without the presence of insects, a chile plant will self-pollinate, a process often desired for breeding purposes. If a breeder wants to grow hundreds of varieties close together, as in the experimental garden, the genetic purity of the plants must be ensured. If seed is to be saved, the flowers of a plant cannot be allowed to cross-pollinate.

"With pure seed, we maintain genetic purity," Bosland says. "What we're doing here is trying to get pure seed, seed that we know is true to type."

The self-pollination process began in late winter, when the seeds of these plants were sown in trays in the experimental greenhouse. In late spring, the plants were transplanted to the field and left to grow. Now that flowers have started to form, Bosland's students are stripping them from the plants, along with any early fruit that has set, encouraging the plants to pollinate themselves. Draping the plants with insect-proof cloth ensures that any fruit that sets under the nylon cages is from self-pollinated flowers and

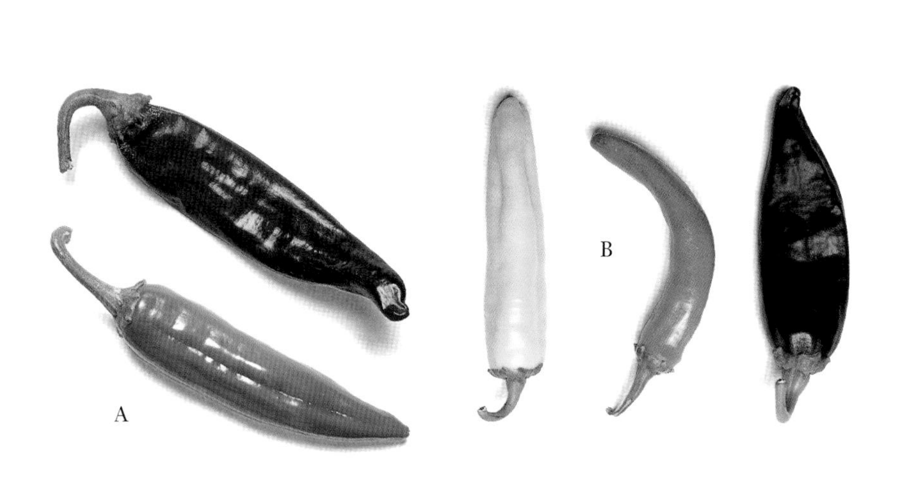

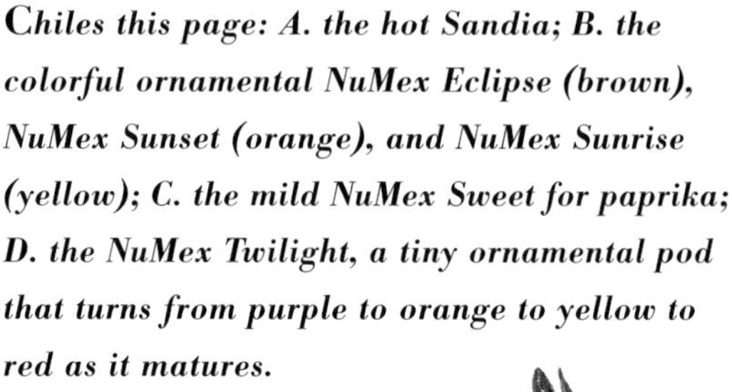

Chiles this page: A. the hot Sandia; B. the colorful ornamental NuMex Eclipse (brown), NuMex Sunset (orange), and NuMex Sunrise (yellow); C. the mild NuMex Sweet for paprika; D. the NuMex Twilight, a tiny ornamental pod that turns from purple to orange to yellow to red as it matures.

is thus genetically pure. When the pods reach maturity in the fall, the seeds are extracted and saved for further research work. Some seeds also are sent to the U.S. Department of Agriculture's Plant Introduction Station in Georgia, where the country's *Capsicum* collection is held and made available to researchers around the world.

When cross-pollination is needed to make a controlled cross between chile plants, Bosland and his students do it by hand. First, an unopened flower bud is selected. Using forceps or tweezers, the flower petals are carefully removed and the reproductive organs exposed. To prevent the plants from self-pollinating, the anthers, where pollen is produced, are removed. Pollen is then collected from the open flower and anther of another chile plant and, with a small paintbrush or stick, transferred to the stigma of the first plant. Before long, the fruit will set and the chile will mature. Since the germ plasm for the new chile cross resides in the seeds that are produced by the pollination process, the cross will not be seen in this year's pods. Once the chile ripens, the new seeds are saved to be planted the following season, when the characteristics of the new cross will be revealed.

Depending on which specific genetic systems were combined in the cross, new cultivars can exhibit a variety of new traits. The first test of a successful cross is the physical structure of the plant. "The plant needs a certain architecture," Bosland says. "It has to grow a certain way for this style of farming we have here." A plant must grow six to eight inches tall on a strong single stem to prevent it from touching the ground and thus rotting. The plant's stem should begin to branch up high, and because it is advantageous for a larger percentage of a plant's early flowers to set fruit than its later flowers, the plant must be designed so that the first flowers are not dropped.

Once the plant's architecture is established, Bosland can concentrate on other equally important things, such as substituting Easter chiles for Easter lilies or breeding chile that improves the feather color of flamingos in zoos.

The word *Capsicum* is derived from the Greek word *kapto*, "to bite." After color and shape, chile's most recognizable trait is its bite or, more precisely, its pungency. For the average chile eater, the level of heat in chile often depends on whether the label on the salsa jar says mild, medium, hot, or extra-hot. From the scientific perspective, though, heat is determined by a combination of the seven closely related chemical compounds called capsaicinoids that are found in no other plant. Capsaicinoids are produced in glands on the placenta, or vein, of the fruit. In most chiles, capsaicin and dihydrocapsaicin are the two capsaicinoids that generally appear in the greatest amounts, although each of the seven compounds

The Heat of the Matter

Keeping up with the chile industry also means keeping up with the latest technology. For researcher Paul Bosland, some of the most important technological strides made in chile research in recent years are embodied in the High Performance Liquid Chromatograph (HPLC), a state-of-the-art pungency detector. Its precursor, the Scoville Organoleptic Test, was the first laboratory test used to measure chile pungency and has, since 1912, been the standard systematic approach for determining the amount of heat in a chile pod. The test relies on human taste buds to detect pungency in solutions of one part per million. In this method, trained tasters evaluate a chile sample and record its level of heat. The samples are diluted until the testers can no longer detect heat. The dilution is called the Scoville Heat Unit.

Although the Scoville method is still appropriate in some circumstances, the human palate generally provides a much too subjective measure of chile heat for research purposes. The HPLC solves the subjectivity problem with more consistent, objective, and reliable results. With this method, the chile pods are dried and ground to a fine powder. The capsaicinoids, the chemical compounds that promote pungency, are then extracted from the powder and, through a process of mixing and heating, separated out into the seven individual capsaicinoids found in chile. The machine not only measures the total heat of the chile pod but also determines the amount of heat in each individual capsaicinoid, valuable information for pharmaceutical use.

The HPLC measures pure capsaicin in parts per million. Researchers convert these measurements to Scoville Heat Units by multiplying them by 15. Although the Scoville taste method has become virtually obsolete by the HPLC, the Scoville Heat Unit is still the accepted industry measure of heat. Humans have a natural pungency threshold of 150 Scoville units, leaving breeders to strike within the range of 300 to 700 Scovilles for low-pungency paprika; 700 to 1,000 Scovilles for mild chile; 1,000 to 1,500 Scovilles for medium chile; and between 2,000 and 3,000 Scovilles for hot to extra-hot chile. Bosland refers to the range between 3,000 and 60,000 Scovilles as "No Man's Land." Beyond "No Man's Land" is the habanero, one of the world's hottest chiles, which hits the charts somewhere between 200,000 to 300,000 Scovilles.

The seeds of every chile cultivar ever grown at New Mexico State University are kept under lock and key in a small, temperature-controlled room. The room is a repository for literally millions of seeds catalogued into some twenty-five thousand distinct chile types.

At $35 thousand per machine, the HPLC is a costly investment, but researchers consider it the most accurate way to measure chile pungency. Such pinpoint accuracy is crucial, Bosland says, as the demand and the competition for chile continue to grow. He cites the emerging strength of Brazil as a red chile producer, as well as the quality chile crops coming out of countries such as South Africa, Korea, China, and Mexico. New Mexico has established a respectable niche in green and red chile, paprika, jalapeño, and cayenne, but Bosland points out that it is a niche that researchers armed with better data potentially could move into.

"It's quaint to look at the New Mexico picture, but you also have to step back and look at the big picture," Bosland says. "If we were to quit this program at the university, somebody else could catch up with us in as little as three years."

Paprika

Nearly one-third of New Mexico's annual chile crop is comprised of cultivars that will be turned into paprika. When it comes in the form of paprika, chile is the number-one natural food coloring agent in the United States. In Europe, paprika, the Hungarian word for chile, refers to one of two specific chile types, but in the United States, paprika is not an actual pod type; it is any low-pungency, powdered chile with a brilliant red hue that can come from any one of a variety of Capsicum annuum cultivars. Among the most popular chiles with New Mexico growers for paprika production are "NuMex R Naky," "NuMex Conquistador," and "NuMex Sweet," all of which were developed at New Mexico State University.

The natural dyes that are extracted from these mild cultivars, in the form of a red oily substance called oleoresin, are a popular coloring agent that are added to a wide variety of foods, including salad dressings, tomato sauce, soups, margarine, cheese, ice cream, hot dogs, sausages, and prepared Oriental and Mexican foods. Blushes, eye shadows, nail polishes, lipsticks, and other cosmetics are colored with paprika-derived dyes, as are pet foods and various pharmaceuticals. Paprika is even fed to captive flamingos to keep their plumage pink.

generates heat. Their presence can make the difference between a pleasant edible experience or one that feels like it came straight from the fiery depths of hell.

"The terms mild, medium, and hot have no meaning to a chile breeder. A company decides what's mild, medium, or hot depending on how it wants to sell a product," Bosland says. "A breeder is interested in the genes that will turn the heat up in a pod that needs to be hotter or make the pungency go to zero for paprika."

The capsaicinoid content of a chile plant partially depends on its genetic makeup. Nonpungency, as in a bell pepper, is determined by a single gene, while varying heat levels depend on the configuration of many genes. Although the control genetics have over pungency is not yet fully understood by breeders, Bosland can manipulate these genes to produce plants within certain relative heat levels desired by consumers. The NuMex Joe E. Parker chile that Bosland released in 1990, for example, was genetically selected to produce medium-pungency pods.

Originally, pungency in the chile plant evolved not in response to the taste buds of humans but of other mammals. Because birds had a natural tolerance for the original wild chile peppers, they became an ally of the chile plant, depended upon to spread its seed around. However, the plants developed higher levels of heat to defend themselves against consumption by certain undesirable mammals. Although many plants in the nightshade family contain alkaloids toxic to many mammals, the levels of capsaicin, the most prevalent alkaloid in the chile plant, are not high enough to be toxic.

Humans, meanwhile, vary in their sensitivity to chile heat. When a person bites into a pungent batch of chile, the capsaicinoids irritate pain receptors located in the mouth, nose, and stomach to create the physical sensation of heat. Stimulation of the sensory receptors results in the release of "Substance P," a chemical messenger that tells the brain there is pain. In turn, the brain floods the nerve endings with endorphins, the body's natural painkillers. Repeated chile consumption will eventually desensitize the pain receptors, which is why some individuals can tolerate very hot foods while others cannot.

Not only is the level of pungency detected differently by the human palate but each of the seven capsaicinoids is perceived differently by each individual. Some capsaicinoids produce a heat that disappears quickly while others have a lingering heat. Some go for the front of the mouth while others aim for the roof, the lips, the throat, or the tongue. In breeding for various degrees of pungency, Bosland considers whether a fast-dissipating heat or a lingering heat is desired. For instance, if the chile is to be used for medicinal purposes, a deep, penetrating heat is key, but for food purposes, he strives for a chile that produces a rapid burn that disappears quickly enough to make an individual willing to eat more.

Still, the primary determinant of pungency is completely out of Bosland's control. As high as 76 percent of the pungency differences in chile is the result of a plant's interaction with the environment. While a grower may select a certain variety of chile for its desired level of pungency, the degree of heat the plant actually achieves will depend on where and how the chile is grown. Factors such as water, temperature, and weather can all have influence. A chile crop that is grown during a cool season will produce milder peppers while one grown under intense heat will result in hotter pods. Although some chiles are more sensitive to the elements than others, any type of stress in the field will make a chile hotter, even those that have been bred for low levels of pungency. Thus, Bosland's challenge is to breed a "stress-free chile," one whose heat stays uniform no matter how hot, cold, or wet it gets.

Pungency has its place in other areas besides food. Oceanographers recommend a fifty-fifty mixture of hot chile powder and petroleum jelly as a preventative for "biofouling," the technical term for the growth of barnacles and other marine organisms on the bottoms of boats and other ocean research equipment. Chile puts the pleasant bite in ginger ale and the not-so-pleasant sting in the popular antimugger aerosols, or chile pepper defense sprays, that many carry for protection. Capsaicin is the powerful active ingredient that causes attackers to gasp for air, twitch helplessly, and be temporarily blinded upon being sprayed. The aerosols have gradually replaced Mace® and tear gas in police departments throughout the United States.

However, it is the medicinal value of chile and its pungent properties that have perhaps generated the greatest interest in the plant, both historically and in recent years. The Aztecs of Mexico knew the curative properties of chile well, using it to stimulate the appetite, encourage bowel movements, aid digestion, comfort stomachaches, and strengthen a system depleted by a cold. After the Spanish conquest of Mexico, the Spanish physician Nicholas Monardes also advocated the medicinal benefits of chile in one of the earliest books on the properties of New World plants, translated into English in 1577. He wrote of the chile pepper: "It dooeth com-

Ornamentals

Breeder Paul Bosland believes he can give New Mexico chile growers a major competitive edge with the development of ornamental chiles. Ornamentals are bred not for taste but for show. From brilliant orange to yellow to purple to brown, ornamentals come in all the shades of the rainbow, often displaying up to five different colored pods simultaneously on the same plant. The chiles are used to make wreaths, ristras, and other popular chile decorations.

Bosland began testing for ornamental types shortly after arriving at the university. In reading the early chile studies of his predecessors, he learned that potted red and green chiles once were popular Christmas gifts known as Christmas peppers. In time, however, the peppers were overtaken by the poinsettia plant as the holiday favorite. Bosland now believes that multicolored ornamentals can again fill a major space in the marketplace.

Since pungency levels and flavor are not considerations in the ornamental breeding process, Bosland concentrates instead on developing such visual characteristics as color and shape. In 1988 he began releasing a variety of new ornamental cultivars. The bright yellow "NuMex Sunrise," the blazing orange "NuMex Sunset," and the dark brown "NuMex Eclipse" were the first standard-sized pods introduced by Bosland that year. The following year, in honor of the one-hundred-year anniversary of New Mexico State University, he unveiled the "NuMex Centennial," a potted piquin-type pepper with purple foliage and berry-sized fruit that turns from purple to yellow to orange

(continued on page 22)

to red as it ripens. The "NuMex Sunglo," "NuMex Sunflare," and the "NuMex Sunburst" were among the bright new ornamentals introduced by Bosland in 1991. The "NuMex Twilight," another tiny, piquin-type pod, also was released that same year and is an example of how even the subtlest differences can be bred into a chile plant. The purple Twilight fruit matures in the same shades as the Centennial plant except that the yellow stage is much more pronounced.

The ornamentals that Bosland develops are ideal for greenhouse cultivation, but most also are suitable for cultivation in a formal garden bed. In recent years, ornamentals have slowly become a strong alternative crop for many of New Mexico's small farmers. The "NuMex Mirasol" ornamental, which Bosland released in 1993, was developed at the request of some of those farmers who asked for a small "Mirasol-type" chile that could be used to fashion chile wreaths. In Spanish, the word Mirasol means "looking at the sun," and Mirasol-type chiles grow in such a way that they appear to be pointing toward the sun. The cone-shaped Mirasol fruit grows in erect clusters of four to six pods each, appearing like petals on brilliant red flower blooms.

Accepting that the Christmas market has been cornered by the poinsettia plant, Bosland currently is breeding ornamentals with other holidays in mind. A pot of pastel pink, purple, and white chiles could one day be an alternative to the traditional Easter lily. And a black chile that turns orange when mature could replace the jack-o'-lantern at Halloween.

forte muche, it dooeth dissolve windes, it is good for the breast, and for theim that bee colde of complexion: it dooeth heale and comforte, strengthenyng the principall members." On the downside, Monardes added that chile could "facilitate the expulsion of gas."

The tremendous nutritional content of chile makes it a powerful preventative medicine. Just one medium-sized New Mexican green chile pod contains up to six times the vitamin C of a Valencia orange, providing 130 percent of the recommended daily allowance plus two grams of essential dietary fiber. Chile also contains the provitamins that the human liver turns into vitamin A. As a green chile pod turns red, vitamin A content increases until it contains twice the vitamin A of a carrot. Eating a half-tablespoon of red chile powder meets the recommended daily requirement for vitamin A, and some studies indicate that both the vitamin A and the beta carotene contained in chile can help reduce the risk of cancer. In addition to the rich flavor of chile, it is low in calories and has virtually no fat.

In New Mexico, eating a bowl of hot chile has long been a traditional remedy for treating a sore throat or a cold because it stimulates perspiration and sinus drainage. The endorphins released by the brain when capsaicinoids enter the human system also provide the body with a sense of pleasure that masks pain. Today, the pharmaceutical industry counts on the lingering, penetrating heat produced by some capsaicinoids to create a host of medicines that heal and soothe. Capsaicin is often the active ingredient in heat-inducing balms used to treat sore muscles, as well as in topical creams used to soothe shingles. Capsaicin-rich medications also are prescribed to treat migraine headaches and for the temporary relief of rheumatoid arthritis. Surprisingly, despite such advances, researchers are only beginning to explore chile's medicinal uses.

Naming a chile is the final step in the chile breeding process. Before a name can be assigned, three years' worth of data must prove that the new cultivar is "stable" or "uniform," meaning that the chile can be consistently reproduced generation after generation. Paul Bosland also has to prove that the new chile has a trait considered unique compared to what farmers are already growing and that the new cultivar is better than existing chile types in terms of its color, disease resistance, yields, or in some other way. If the chile meets all of those criteria, Bosland and a committee of university and state agriculture officials give the pod a name. Whatever the choice, each name is preceded by a "NuMex" moniker to indicate that the chile was developed at New Mexico State University.

Hidden within the string of New Mexico pod names is the evolution of the plant and the Chile Pepper Breeding Program. From the earliest days of breeding come New Mexico No. 9 and New Mexico No. 6, days when only numbers were assigned. Sandia and Rio Grande 21 represented the increasingly broad regional importance of chile in the state from the mid-1950s to the mid-1960s. Spanning the decade between 1975 and 1985, Roy Nakayama's NuMex Big Jim and NuMex R Naky showed the breeder's personal attachments to the crop. Nakayama named the Big Jim after Hatch farmer Jim Lytle, Sr., shortly after Lytle's death. R Naky was named for Rose, Nakayama's wife.

Bosland doesn't want a chile named after himself. "It's considered an insult if a chile is named after you while you're still actively researching because it means you're through," he says. But the names of his cultivars do reflect personal areas of interest. Names such as "Twilight," "Sunglo," and "Eclipse" express the natural brilliance of his multicolored ornamentals, and his most recent green chile release, the 1990 NuMex Joe E. Parker, expresses his gratitude to the field man from the Old El Paso company who brought him growers' requests for a meaty green chile with midrange pungency that is easy to peel. Not only did the Parker pod satisfy New Mexico growers' needs, it has become the number one green chile grown in Mexico as well.

New Mexico State University horticulture students assist in the chile breeding process by stripping a field of young chile plants of early flower blooms and fruit.

If approved, Bosland's latest project should make a lot of chile cooks happy, too. The "NuMex Stuffer" is an extrawide green chile that should support even fatter chiles rellenos than the popular Big Jim pod. The Stuffer is so wide that it won't fit into a can. It will, however, fit into the plastic containers that frozen chile companies use. Bosland's goal is to see that every chile that comes out of his university finds its niche in the marketplace.

No matter how advanced the technology or how slick the marketing slogan, the multimillion-dollar New Mexico chile industry still boils down to a seed.

For all the genetic manipulation that breeders such as Paul Bosland perform, it is the seed itself that ultimately determines what kind of chile will live in the world. At NMSU, literally millions of chile seeds are kept under lock and key in a temperature-controlled vault, ready for research purposes.

The offspring of the university's seed treasury are as diverse as any human population, with their personalities ranging from the heavyset Big Jim, the mild-mannered R Naky, and the reliable New Mexico 6-4 to the plump Joe E. Parker, the fiery Sandia, and the perky "Bailey Piquin." Just as scientists study the human gene pool for clues about human physiology and behavior, Bosland one day wants to have enough seed to represent the world's entire *Capsicum* gene pool. The university's seed repository could then become a resource for chile researchers from around the world who want to know how and why chiles grow the way they do.

A *horticulture student uses a teaspoon to water chile seedlings.*

"Chile is so much more than just enchiladas. It's really a fascinating crop, and there is so much more to do and learn," Bosland says. "We can use chile to teach about biodiversity and the basic sciences. We can learn chemistry by learning about pungency, learn physics by learning about chile colors, learn economics by learning about the marketplace."

In an effort to increase the school's visibility as a chile research center, Bosland founded the Chile Pepper Institute, a nonprofit international organization devoted to the study of chile, at NMSU in 1990. Since then, the institute has become a clearinghouse for chile publications and information for researchers worldwide. Now its success has inspired members to embark on a much greater enterprise: the establishment of the International Center for Chile Pepper, a $2.5-million museum, botanical garden, chile archive, and research facility that will bring the university's years of knowledge on the subject of chile to the public. Besides its importance as an academic research center, Bosland envisions the center as a major tourist attraction. As more people learn about chile, he believes, the demand for the product from New Mexico growers will increase as well.

"What I love to do is focus on those things that have the most impact, short term and long term, on the industry," Bosland says. "You can't just be a chile breeder; you have to be a bit of a dreamer, too."

Every year, New Mexico State University's one-acre experimental garden is home to as many as one thousand different varieties of chile from around the world.

The Color of Money

Dickie Ogaz has made paprika cultivars the primary focus of his annual chile crop. Opposite: Chile season on Dickie Ogaz's farm is a season of striking colors and contrasts.

Most of the chile grown for paprika production in the United States is grown in New Mexico. New Mexico also exports paprika for use as a natural dye to Africa and Europe, regions that once dominated the world market in paprika production. The popularity of the product has given farmers like Richard ("Dickie") Ogaz, of the southern New Mexico town of Derry, a significant new edge in the competitive chile marketplace. Since Dickie began growing paprika cultivars in 1990, the colorful pods have become the bulk of his annual chile crop.

"I'm not much of a gambler; I don't play the market," Dickie says. "Before I plant, I have it all contracted out. I have a plan for practically every pod."

Farmers who grow chile for paprika earn more per pound for the product than for the pungent edible red. Success, however, does not depend on the size of a pod but on the intensity of its color. The market value of a crop is determined by the surface color of the chile as well as the levels of carotenoids, or color pigments, that can be extracted from the pods. Extractable color is measured by a method developed by the American Spice Trade Association and expressed in "ASTA" units. The higher the ASTA color value, the brighter the product created from the chile's natural dyes.

In order to achieve the brightest possible color, Dickie lets his paprika cultivars ripen in the field much longer than the red chile he grows for culinary use, waiting until about December to harvest the pods. By the time the pods reach their peak, their skins will be 50 percent dry and the color of ripe strawberries.

"I like red. That's why I grow red chile," Dickie says. "Every day, I tell my wife, 'I'm gonna go walk the fields.' I walk up and down the chile rows. It puts me in a different mood."

JOHN DEERE

[Chapter Three]

The Chile Landscape: North and South

A new crop of chile emerges in a field in southern New Mexico. Opposite: The Rio Grande takes on many faces as it winds from the northern to the southern parts of the state (clockwise, top left), irrigating thousands of acres of chile along the way.

A dirt road meanders along a nameless hill in northern New Mexico. Just over the hilltop, where the road straightens and the land turns flat again, an unpicked field of chile comes into view. It is early October. The chile pods hang red and crowded on their stands, surrounded by leaves that are beginning to wither and brown. The midmorning sun casts an unusual texture upon the chile field. From a distance, contrasting flecks of red and brown blend together to look like a speckled shag carpet that has been laid over the earth. But up close, the plants, with their sun-dried fruit slowly dropping from their foliage, appear to have fallen victim to sheer isolation.

The road goes on and the isolation grows greater until, just beyond a bend lined with gold-leafed cottonwoods, a compact adobe house appears. Inside the kitchen a woman, a farmer, takes slow, deliberate sips from her coffee cup. The chile field on the hill has not been forgotten. With two other chile fields to harvest on her own, the woman can only hope that the third field—and the fast-approaching winter frost—will wait.

Some 350 miles south, another farmer drinks coffee in the front seat of a blue Jeep. She looks out the window upon a chile field so large that one can hardly see from end to end. There, nearly a hundred chile pickers are bent low between the red chile rows, their straw hats and bandannas barely visible above the waist-high plants. Pod by pod by pod, the workers fill their buckets, occasionally standing to dump the chile into large wooden bins that sit on the back of flatbeds in the middle of the field. For every bucket that goes into the bins, the workers collect a plastic coin exchangeable for cash at the end of the day.

Whether it is sitting in a field waiting to be picked or hanging on a string from the side of an old pickup, chile can be found throughout the New Mexico landscape. Opposite: Located at 6,500 feet in the northern village of El Guique, Ranch O Casados Farms provides a majestic view of the snowcapped Sangre de Cristo Mountains.

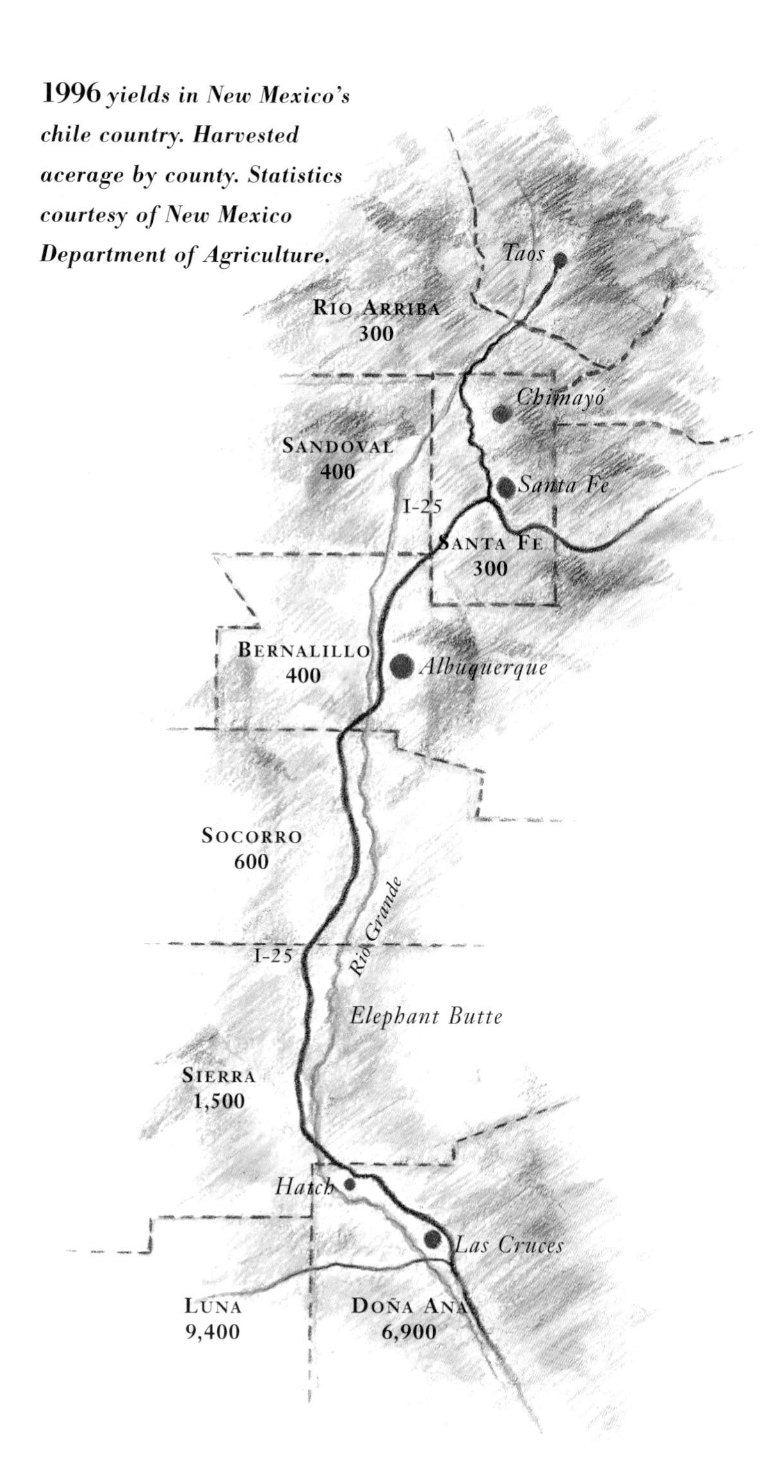

1996 yields in New Mexico's chile country. Harvested acerage by county. Statistics courtesy of New Mexico Department of Agriculture.

The workers pick quickly, though their motivations are different from those of the farmer up north. In this sunbaked section of south-central New Mexico, winter arrives about December, if it decides to come at all. That leaves at least two more months and hundreds more acres of red chile to gather.

There are approximately forty-four million acres of farmland in New Mexico and nearly fourteen thousand farms. Of that land, an average of thirty thousand acres is devoted to chile each year, grown among an estimated 250 to 300 farmers across the state. Since 1990, those farmers have consistently made chile one of the two most profitable crops in New Mexico. With the exception of 1992, a record year for chile production in the state, chile was second only to hay most years in cash receipts. In 1992, the crop was valued at $67.3 million straight off the farm. After processing, the value increased nearly fourfold, exceeding $250 million.

Each year, more than half of the state's chile is grown in the southern counties of Doña Ana and Luna. About seven hundred acres of chile, with an estimated market value of $1.5 million, are cultivated in the northern counties of Rio Arriba and Santa Fe. The overall production of chile is largely determined by region. In northern New Mexico, for instance, the growing season is usually only half as long as in the more temperate south. The size and market value of a farmer's crops thus depend as much upon the landscape as upon the grower's expertise.

The farmer who tends her isolated chile fields in the north grows mostly to feed her family. What she has left, she will sell at a roadside stand or at the area farmer's market from the bed of her truck. The farmer who oversees the large fields in the south grows principally to fulfill contracts with hot-sauce makers in Louisiana.

New Mexico's climate is optimum for growing vast yields of high-quality chile. Irrigation is essential to the process for farmers across the state as precipitation during growing season seldom exceeds seven inches a year. A chile pod grown in northern New Mexico has a flat, or square, shoulder separating the chile from the stem. A southern New Mexico chile pod's shoulder is sloping, almost round. Beyond this it is impossible to generalize. Wherever chile grows, the soil conditions are so diverse, the geographic contrasts so extreme, that literally every single pod grows in its own way, at its own pace.

Perhaps the only thing one can say about chile in New Mexico is that it grows successfully virtually everywhere, at practically every bend, bump, and rise in the road. Chile grows as green at the base of a remote Abiquiú canyon as in an open field in the fertile southern valley. It turns red beneath the same golden sun and azure sky. From cottage industry in the northern villages to agribusiness in the vast farms in the south, the cultivation of chile brings the scattered, independent segments of New Mexico together as a whole.

A string of red chiles, their crooked stems pinned to a crooked clothesline, has been left to dry in the yard of a Chimayó home. The sun reflects off the steep pitch of the tin-roofed house, hitting the glass of a rearview mirror that juts out of the driver's side of an old truck whose front end rests on cinder blocks in the dirt drive. A small chile ristra, half-red, half-green, hangs from the mirror. A sign, "Chimayó Holy Chiles," is posted on a fence nearby.

To a farmer anywhere in New Mexico, faith in nature is the ultimate reason to farm. In the northern village of Chimayó, faith has as much to do with the survival of the community itself as with sustaining local farmers. The small Hispanic farming community is also one of the most important religious sites in the New World. Countless thousands of pious pilgrims are lured to the legendary Santuario de Chimayó each year, hoping to find miracles in a handful of holy earth. In the fall, many return on a pilgrimage of a different sort: to sample the year's chile crop.

Unlike the hybridized chile plants commonly cultivated in southern New Mexico, Chimayó chiles are considered land races, the scions of the first chile seeds planted by the Spanish in the area. Small and crooked with a skin as thin as the outer layer of an onion, Chimayó pods are known for their subtly sweet flavor and mild-to-spicy heat. In the early part of the century, chile was Chimayó's most important food crop. Dried Chimayó chile was then so valuable that it was an acceptable replacement for cash at the area mercantile. Today, Chimayó chile cannot be exchanged at area stores, though its value may be greater than ever before. As in much of northern New Mexico, the older generation of Chimayó farmers can find few young villagers to pass down their chile seed to. Most residents have abandoned the crop completely. The future of chile here is bleak.

Nonetheless, driving past the "Chimayó Holy Chiles" sign, along the winding road that accesses the far-flung houses of Chimayó, one arrives at the home of Gonzalo and Ermenda Martinez, who grow authentic Chimayó chile on a half-acre plot ringed by giant sunflowers. The chile they grow is sold at the summer farmer's market in Santa Fe, but more than for profit these two Chimayó natives grow to preserve the quality of life their ancestors knew. It is a way of life the two have practiced together since their marriage in 1950, when Gonzalo was nineteen and Ermenda thirteen. From their spacious adobe home, which is reached by an out-of-the-way arroyo that doubles as a county road at the western edge of Chimayó, the couple has sustained the land-based life-style that once defined the villages of northern New Mexico.

Chile is dried and a fresh harvest of corn is sorted in Chimayó, ca. 1950. Photograph by T. Harmon Parkhurst. Museum of New Mexico Neg. No. 40362.

"My mother always used to say, 'If you plant it with joy, it will grow,'" Ermenda shares, as she pulls four jars of homemade jam, two apricot and two peach, from the cupboard and sets cups of coffee on the wooden tabletop. She opens one jar and arranges some bread and butter on a plate; the other jars she offers to her visitors as gifts. "My mother also used to say, 'If somebody comes to your house, always take them inside and feed them, and you will never run out of food.'"

Ermenda grew up harvesting a variety of fruits and vegetables on land that her parents farmed in Chimayó. "My parents didn't have any land of their own," she recalls. "My stepfather had to plant half the land to give to the owner and the other half was for our family to eat." Gonzalo's father, Melecio Martinez, owned ten acres around the Martinezes' present home. When Gonzalo and Ermenda married, Melecio gave them their own parcel of land. They, in turn, eventually divided a portion of their land among each of their four children, all of whom still live in Chimayó.

It is late summer and the couple's front porch is crowded with cardboard boxes of crimson tomatoes and green Chimayó chile picked to sell at the market the next day. Just beyond the porch, a row of corncobs has been husked and strung across the driveway to dry in the hot summer air, their exposed kernels covered with netting to keep the birds away. The corn will be sold in its dried form as *chicos*, hard, sweet, crunchy kernels often added to beans or made into stews of their own.

"We used to plant that acre across the road, too," Gonzalo says, indicating east with his bearded chin as he bends to pet one of the various dogs that wander about the yard. "In 1980, we gave that land to our daughter."

That left Gonzalo and Ermenda with a half-acre of earth; a garden, they call it, not a farm. Every year, they plant chile, tomatoes, cucumbers, melons, okra, radishes, and sugarcane, sprinkling various flowers into the mix until a rainbow of sunflowers, dahlias, hibiscus, daisies, and marigolds fills in the spaces between the vegetables and the fruit. Like many northern New Mexico farms, the aesthetics of the garden is as important as its productivity.

"It's beautiful," Ermenda says as she looks upon the fertile field. "Being surrounded by beauty is what makes us want to work so hard out here."

The couple walks to the far end of the garden, where about ten rows of chile grow small and crooked with paper-thin skins. The chiles would look like dwarfs alongside a giant Big Jim, but these are the distinctive pods for which Chimayó is known. Because of their smaller size, Chimayó pods are not as meaty as other New Mexico chile types, but northerners swear by their sweet heat.

"It's small but it's got a kick," Gonzalo says after biting into a tiny, tongue-stinging pod. "Chimayó chile has a unique flavor. The taste just kind of crawls into your throat." Indeed, of all the chile-growing areas in northern New Mexico, Chimayó enjoys a particularly distinguished reputation for its crop. Chimayó is the region's most recognizable name in chile.

Like the Santuario de Chimayó that is at the heart of the community's

Top: To many in Chimayó, site of the legendary Santuario de Chimayó, chile is as sacred as the church's holy earth. Above: The Oratorio de San Buenaventura was built in the early 1900s in Chimayó's Plaza del Cerro.

Since it was built in 1816, the Santuario de Chimayó has lured thousands of religious pilgrims to Chimayó.

religious life, chile has been central to Chimayó's social culture for centuries. Reports of chile being grown in northern New Mexico date to 1598, shortly after the arrival of the Don Juan de Oñate expedition in the region. In time, numerous specialized varieties of chile developed as distinct cultivars and were named for the areas in which they are grown: Dixon, Velarde, and Chimayó.

Situated at the head of the Santa Cruz River Valley, Chimayó was settled shortly after the Spanish reconquest of the region in 1692, taking its name from Tsimayo, a Tewa Indian pueblo once located there. Like most other northern New Mexico villages, Chimayó began as a community of farmers. During the nearly 250 years between the reconquest and World War II, Chimayosos grew fruits, vegetables, and herbs for family use. As Chimayó chile seeds were passed down through generations of farmers and as countless chile crops were reaped and sowed, the Chimayó chile pod evolved into a fruit that reflected the climate, soil, and other conditions in which it was grown. The pods' diminutive proportions and often shriveled surfaces were a natural adaptation to the short growing season, rugged terrain, and unpredictable weather patterns characteristic of Chimayó; but while the elements effectively stunted the growth of the pods, they also encouraged the development of an unusually hardy breed of chile known to survive one, two, even three killing frosts.

Of all the village's homegrown products, chile emerged as an indispensable dietary staple. Along with *quelites*, or wild spinach, chile was one of the first vegetables widely grown in Chimayó, and almost every resident had a *huerta*, or chile field, of his own. A well-defined division of labor was observed during the production of the crop. The job of preparing the land for the crop was held by the village men, while the major tasks of cultivation—planting, thinning, weeding, irrigating, and picking—were the women's domain.

Like many other locally grown herbs and plants, villagers exploited the medicinal qualities of the chile plant, eating extrahot chile to cure colds and sore throats. But ill or not, Chimayó residents ate chile at practically every meal of the day. They roasted and peeled its thin skin to eat it fresh and green or tied it into ristras to turn red and dry. The time-consuming task of stringing chile into ristras was a social activity in which family and neighbors took part, sharing local news and gossip while they tied. They then hung the chile strings from porches, vigas, and building facades to dry in the sun. Later, residents hoisted gunnysacks full of the dried chile upon their backs and carried them to a local *molino*, or gristmill. There, the pods were pulverized into a fine powder or crushed into a coarser *caribe* grind. The chile ultimately would be cooked into savory sauces to be eaten alone or in combination with other foods.

Every fall, farmers ground enough chile to carry their families through the winter, commonly exchanging a portion of their product as payment to the owner of the mill. This friendly trade between neighbors eventually gave way to a sophisticated barter system centered around the annual chile crop. As the village's primary cash crop, chile did not necessarily bring cash to farmers in the way a truly commercial crop would, but its use as a cash substitute provided an economic base for the community and a means of obtaining goods that villagers otherwise may not have had. For this reason, chile was the only crop that Chimayosos regularly grew in excess of their personal needs. At harvesttime, farmers loaded their surplus ristras into horse-drawn wagons to take to neighboring communities for trade. In nearby northern villages such as Peñasco, Truchas, and Taos, where the plant was especially difficult to grow, chile was exchanged for wheat, fruit, corn, and other crops. In Mora, to the east, goat cheese and mutton were the most common trades.

The arrival of the narrow-gauge Denver and Rio Grande Western Railroad in northern New Mexico in 1881 substantially increased the trade value of chile while expanding the area of exchange. Along with apples, piñon, and Chimayó blankets woven from hand-spun wool, ristras were shipped north along the Chili Line from Chimayó to the San Luis Valley of southern Colorado. The train back from the valley brought beans, potatoes, and sheep for wool.

By the early twentieth century, dried Chimayó chile had become so valuable that it was an acceptable replacement for cash at mercantiles in Española and Santa Fe, where villagers used it to buy everything from food and clothing to equipment for the farm. The barter system was now so specifically geared to chile that items were priced not only by the dollar but by the ristra as well. A sack of potatoes was an even exchange for one-and-a-half chile strings. Two ristras could be traded for either 140 pounds of wheat or sixteen pounds of beans. But World War I would change the prevailing system of trade from barter to a cash basis so that farmers received only store credit, not cash, for the product. Each ristra was carefully inspected before it was accepted by a mercantile store, and prices were assigned according to whether the ristra was deemed a first-, second-, or third-grade string. A first-grade ristra had to be at least five feet long, tightly strung, with a minimum of rotten pods. At Bond and Nohl mercantile in Española in 1930, a first-grade string was equivalent to $1 in cash. With the onset of the Great Depression, the price per string plummeted to as low as thirty-five cents, but with employment scarce throughout the region the crop's barter value was perhaps more precious than ever before. In 1935, one-third of all the land in cultivation in Chimayó was devoted to chile, and some farmers produced up to 250 ristras per acre. By 1937, chile remained the villagers' principal means of obtaining clothing, flour, lard, sugar, coffee, and beans.

Chile also retained its culinary value in distant markets during the depression. Northern New Mexico chile growers continued to command a faithful market despite the fact that chile growers from California, Texas, and southern New Mexico consistently priced their chile cheaper. Throughout the 1930s, some sixty thousand ristras were produced annually in Chimayó and other northern New Mexico villages for trade at area mercantiles. Bond and Nohl bought approximately fifteen thousand ristras a year, shipping either full strings of chile or the powder they produced to destinations throughout the West. In 1937, 65 percent of the chile produced in the region was handled by such merchants. The other 35 percent was sold by the farmers themselves, who now used automobiles to transport their chile to neighboring towns to sell to private individuals, small grocers, and restaurateurs.

According to local legend, up to 150,000 pounds of chile were grown annually in Chimayó when the crop was at its peak in the late nineteenth and early twentieth centuries. But with the drain of a growing population on the village's limited land resources, the economic stresses of the depression, and the start of World War II, the traditional agrarian life-style of northern New Mexico began to wane. Many returning home after the war looked to nearby Los Alamos National Laboratory, the secret scientific community, for jobs and the promise of financial security at a time of limited economic

opportunity. While many old-time farmers continued to supplement their incomes with garden-size chile crops, the prospect of passing their chile seeds down to the next generation became more remote as the declining economic opportunities in the region during the postwar era forced more people to move away.

In the fifty years since, the production of chile in Chimayó has steadily declined. Although chile is marketed today under the Chimayó name, there is scarcely enough chile acreage in the village to fill the demand at area farmer's markets and shops. The situation has prompted some people to sell chile from other places under a false label of "Chimayó."

For Gonzalo and Ermenda Martinez, the erosion of the local chile crop has stirred conflicting feelings of opportunity and loss. As they are among the few local farmers who still maintain the crop, the Martinezes' chile is in greater demand than ever before. Yet knowing that the agricultural tradition of their village is dying has left them with a profound sense of sadness. "We've been growing chile all of our lives," Ermenda says, "but someday there may be no more chile in Chimayó."

For now, the Martinezes say they will do their part to maintain the Chimayó chile tradition by growing chile in their small garden each year and by educating farmer's market customers about the history of the crop. They know it is unrealistic to think that farming could ever again support the economy of their growing village and that the era of dependence on the land is long gone. But someday, perhaps, their grandchildren might develop a genuine interest in the agricultural life their grandparents now share. Maybe one day they might even plant a patch of Chimayó chile all their own.

"You never know," Gonzalo says, a hopeful expression lighting up his face. He looks at Ermenda. She shrugs her shoulders, smiles. "You never know."

Chimayó natives Gonzalo and Ermenda Martinez have been growing authentic Chimayó chile together since they married in 1950.

There are nails protruding from the massive pine vigas that traverse the second-story ceiling of Jody Apple's 1770s Chimayó home, rusty remnants of the red chile ristras that once were hung there to dry. The mid-afternoon sun streams into the spacious room where the original owners stored chile, grains, and other goods. No chile dangles overhead now; instead, the sun illuminates a small kitchen whose counters are lined in colorful Mexican tiles and a sitting area appointed with a cozy couch and chairs, a fireplace, and a color TV.

Jody purchased the sprawling eight-thousand-square-foot adobe house and its surrounding four-and-a-half-acre property from descendants of the original owners in 1989. Formerly from Los Angeles, where she ran a children's clothing store, Jody moved to New Mexico in 1982. She lived in Taos for a time before Chimayó's beauty and rural ambience lured her south.

Jody heard of property for sale in Chimayó's historic Plaza del Cerro, a late-eighteenth-century fortified plaza that has survived as the only intact defensive Spanish colonial plaza in New Mexico. Available for purchase was a long adobe building that stretched along the plaza's south side. The site of one of the plaza *torreones*, or defensive towers, the building served as the

Jody Apple (pictured) restored the 1770s Chimayó home and mercantile of the Victor Ortega family back to its original opulence as a residence and working chile farm. At left, a view of the Ortega home before restoration; this page, the property restored.

original residence of one of the two renowned Ortega clans of Chimayó. In the early twentieth century, it belonged to Victor Ortega, who at the time was the most influential person on the plaza: a probate judge, postmaster, school director, and staunch Republican.

Like the rest of Chimayó, the original inhabitants of the Plaza del Cerro grew chile for family use and trade. The Acequia de los Ortegas diverted water from the Rio Quemado above the Chimayó Valley to feed residents' crops. The Ortega ditch, one of the oldest intact acequia systems in New Mexico today, still snaked through the Ortega property when Jody first saw it in 1989, but the old adobe, once the most opulent home in the plaza, stood in woeful ruin. The farm fields were barren, with no traces of the chile crops that once grew there.

Milagro, Jody Apple's gray Appaloosa, was raised alongside Apple's first chile crop.

Jody undertook a full-scale restoration of the building, stressing its historic architectural details and returning it as much as possible to its original character. Today, a series of shiny pitched metal roofs cap the newly plastered adobe facade and slant down toward an emerald-green field below. Jody even restored a small stone-lined pond that the Ortegas had established as a holding pond in case the ditch water ever ran out. In purchasing such a deeply rooted piece of Chimayó history, Jody stirred a sense of nostalgia for the past among her neighbors. At the same time, she dredged up a few old negative passions about newcomers, particularly Anglos, moving to Chimayó.

Nonetheless, Jody found little resistance among local residents regarding her plans to reestablish the Chimayó chile crop that the Ortegas once grew on the property. In 1993, she approached some of them to ask for the seed she needed to start a crop of her own. Her next-door neighbor, Perry Trujillo, offered to introduce her to some of the village's old-time farmers, but it was up to Jody to convince such serious chile-growing veterans that she was sincere about learning to properly cultivate the crop.

"It was like I had to be introduced to all these old-timers to get them to give me their seed," she recalls. "I told them that I think any seed, any plant species, that has survived hundreds of years, that it would just be pitiful to let it go, especially something that tastes so good."

It was Carmen Salazar and Aurelia Vigil who finally gave Jody their generations-old seed, while Ezequial Trujillo, whom Jody affectionately calls "Primo Ezequial," volunteered to teach her to plant. Apple told Perry and Ezequial that she planned to start planting after a brief trip out of town. When she returned, she discovered that the fields already had been plowed by Perry and shaped into planting rows. The two then tested the irrigation ditch and discovered that the water didn't run properly between the furrows. Perry reshaped the rows, and this time, the water flowed smoothly, turning the color of coffee as it soaked into the rich brown Chimayó earth.

Meanwhile, Ezequial patiently waited for the right time to teach Jody to plant. The time finally arrived on a pristine May morning as he instructed her to sow the seeds on the side, rather than the middle, of the planting rows. Ezequial, weathered and brown, faced down the dirt and demonstrated the process himself, emphasizing to Jody that she was to leave eighteen inches between every plant. The old man worked with the easy grace of one who has undertaken a task hundreds of times before. Three hits of the hoe and a handful of seed. Three hits of the hoe and a handful of seed. Over and over, Ezequial repeated the action like a mantra, his own time-honored chile chant. "It was like he was dancing through the field," Jody recalls.

The plants popped out of the soil about a month later. When they stood four inches tall, Ezequial showed Jody how to thin them with a hoe. He told her to fertilize the chile once when small and then left her to tend to the watering and weeding.

Somewhere between the time when the plants flowered and their fruit began to set, Jody heard that Perry's pregnant Appaloosa was dying. The local veterinarian was forced to take the horse's foal but gave Perry little hope for the filly's survival. Jody volunteered to take the motherless pony in, and for three weeks it literally lived inside her house. To get the newborn to nurse, she smeared honey on her face and urged the filly to suck it off. Jody then switched to a bottle and fed the pony by hand. The filly required constant care, leaving Jody less time to attend to her chile crop, but as news of Jody's generosity spread, neighbors pitched in to help. Even some who had viewed her as an unwelcomed newcomer stopped by to water or otherwise tend to the crop.

"One of the nicest things was that all of the neighbors came to see the horse," Jody recalls. "One of them said to me, 'God bless you for saving this horse.' " The filly grew healthy and strong as Jody's first acre of Chimayó chile grew to maturity. She harvested 2,500 pounds from her first crop, and Primo Ezequial showed up to help with the picking. For every bushel the others picked, he picked two and a half. In the meantime, Perry presented Jody with a gift—the horse she had nurtured from birth. Jody named the horse "Milagro" for the miracle of her life.

Today, Milagro has a long wild mane and a white blaze running down the middle of her face. She lives in a pasture adjacent to Jody's chile field, where this early October afternoon two miniature dachshunds are chasing a cat amid the small distinctive chiles of Chimayó. Although most of the pods are short, a few have grown so long that their thin tails are curling up in the air. Jody grew only one-quarter acre of chile in 1995, or about five hundred pounds.

"This was a hard year," she says. "Things didn't really start going until June." The last few days have been hard as well. First, twelve marauding cows went traipsing through her chile field and trampled her plants. Then, a few nights ago, all of her cucumbers and squash froze. Still, like Milagro, Jody's chile plants somehow managed to survive.

"The thing I like most about this chile is that because it's been raised here, it's really been acclimated, which means it's very frost-resistant," she says. "I think if you were to take a seed like this out of its area, it would never be the same. It's the soil, the weather, the water here that make this chile what it is. If you grew it somewhere else, it would change."

Jody bends down and starts to fill a bushel with the scarlet fruit. The smaller size makes the pods harder to pick than other larger types and thus takes her longer to load up a bushelful. "Chimayó chile is an enormous amount of work," she says as the pods begin to pile up. "It's not like growing corn, where you weed it once and then let it go. You have to work for every little pod."

After about twenty minutes, the bushel is brimming. Jody hefts the load onto her left shoulder and carries it across the field and up a set of stairs alongside her house to the former storage room. She places the pods upon window screens that are spread out on the floor. There in the sunlit space, the chile will become even smaller as it shrinks and dries over the next few days. Jody will then package the pods and powder in clear plastic wrap so that their brilliant color shows through. The final product will be sold under the label "Chimayó Chile Company" and marketed to restaurants and shops. Most of this year's crop, however, will be saved for the seed that Jody will use for subsequent crops.

"These guys around here sell this seed for something like a buck a pound," she says. "And with all the expense and the work, who can blame them? Who in the world in their right mind would want to do this?"

Indeed, one wonders why such a woman would choose such a life. Jody has now been accepted in her adopted community to the extent that she is a commissioner on the governing body of the Ortega acequia, a title traditionally held by local Hispanic men, but why would she want to entangle herself in the complex social politics of water? Like Gonzalo and Ermenda Martinez, Jody mourns the decline of the chile tradition here and can see that losing chile means losing something much more profound.

"Obviously, people don't grow enough chile here anymore to make it economically feasible, and that's too bad," Jody says, "but it's not even about money, it's about life. In the old days, people would get together, they'd roast the chile, they'd peel the chile, they'd tie ristras late into the night. I mean, what's happening to the American family that such simple, beautiful things are no longer being done?"

The old man steps carefully down the stairs—step, pause; step, pause; step—his uneven gait supported by a crooked wooden cane. Almost before one can make out the man's small shape, one sees the driftwood walking stick that paces before him. Its tip curves at a perfect angle into the man's right hand, standing at just the right height to balance his bad left hip. The cane looks like a snake that slithered up from the ground and into his palm and then froze in midair.

The driftwood is smooth and brown now, but it was rough and sun-bleached when Orlando Casados found it at Abiquiú Lake many years ago. At the time he had no particular use for the stick, but he liked its shape. He carried it home and tossed it into a barrel on the front porch, where it remained until many years later, when the retired farmer needed some extra stability.

"I got ahold of it, and it felt so good," Casados recalls, sitting now on the living room couch. He painted the stick brown and put a rubber stop on the bottom. "Sometimes life works out that way. The things that are meant for you will wait."

When Orlando Henry Casados, Sr., brought his wife, Marie, and their six children to northern New Mexico from Denver in 1959, he had no idea what awaited them there. He was born in Trinidad, Colorado, where he owned two taverns and sold beer at ten cents a bottle. With the start of World War II, he sold the taverns and moved to Denver to mold defense machinery parts in an iron foundry. It was hot, dirty, greasy work, but Casados worked his way up to foreman with a $1.62 hourly wage. He stayed on at the foundry after the war, making parts for furnaces, stoves, and farm machines. Eventually, he bought some farmland: six acres here, six-and-a-half there, another ten way over there. He grew sweet corn and other vegetables, and come midsummer, he sat his children in a stand beside a busy Denver boulevard to sell the produce to passersby. All Casados had ever wanted was to farm, but there was not enough water in Denver to do so successfully. So in 1959, Casados loaded his family into the car and left the foundry and Colorado behind.

Traveling south through Alamosa and across the New Mexico border, Casados cut east over the Rio Grande and made his first stop in Taos. He liked the look of the town, but after chatting with a few locals he realized that the climate was too cold, the elevation too high, for large-scale farming. A banker in Taos told him to go south toward Española and see the county agricultural agent there. The man had heard of some land for sale in the Española Valley on the western banks of the Rio Grande. He said he didn't think there were many farmers there, except for the Indians who lived at the nearby pueblo of San Juan. Plus, rumor had it there was water galore.

"You came to the right place," the Española county agent said upon meeting Casados the next day.

The agent directed the family seven miles north of Española, through the Tewa-speaking pueblo of San Juan and across a funky, one-lane bridge that straddled the Rio Grande. About two miles north of the pueblo, an obscure little glen called El Guique appeared. Situated beside the big river, in the shadow of the Sangre de Cristo Mountains, the village's name likely was derived from the Tewa language and may well have been the site of an eighteenth-century pueblo. The community had since shifted to one largely populated by Hispanics.

The property owner showed Casados a twelve-acre tract of land with a pink, pitch-roofed adobe house across the road. The house sloped down toward the ancient El Guique Acequia, one of the first irrigation ditches off the Rio Grande. The acequia, which by then was already hundreds of years old, was a plentiful source of water for the land below, as well as for another thirteen acres up the road.

The land was a thick forest of ancient cottonwoods that eclipsed the view of the river below. Somewhere in the middle of the trees, Casados saw something jutting above the ground, a piece of iron railroad track that once

Orlando Casados, Sr., was drawn to the northern village of El Guique by its abundant water. At left, the El Guique Acequia.

Since his father's retirement, Orlando Casados, Jr., has overseen the production of chile and other crops at Ranch O Casados Farms.

was part of the Chili Line railway that traversed a north-south path from Antonito, Colorado, to Santa Fe. He liked the history that was rooted in this landscape and the challenge of its rugged beauty. He paid $25,000 for the property, which came to be called Ranch O Casados.

In the spring of 1960, after clearing the land for cultivation, Casados planted five hundred apple trees. Although the farm was situated a steep 6,500 feet above sea level, the climate in the sun-soaked Española Valley proved just right for apple growing. In addition to apples, Casados grew a small crop of the local chile for his family, but it never occurred to him to follow the suit of his neighbors and market the chile. Apples were to be his cash crop.

A rare cold snap paralyzed northern New Mexico in January 1971, and nighttime temperatures in Española plunged to thirty-three degrees below zero. Casados got up every four hours to start the tractors and other farm machinery so their fluids wouldn't freeze. When his son Orlando went outside one night to start up one of the machines, he didn't consider that he had a hole in his glove. He returned to the house with his finger frostbitten and numb.

"It was bitter, bitter cold," Casados recalls. "In the daytime, the sun was shining, but you'd go outside and even your eyeballs got cold." Five hundred apple trees froze. "That's when we went into chile," he says. "From then on, we went into chile more and more." Casados rises from the couch, catches his balance, and shuffles outdoors. Standing now beside the ten-foot-wide El Guique Acequia, overlooking what has to be one of the most pristine farms in northern New Mexico, he is undoubtedly older, grayer, and slower than he used to be, but now he is even more proud.

"We've got the best water rights in the United States right here," he says. "Our water's clear; you can drink it. The river gets dirty further down where it hooks up with the Rio Chama, but we never, never, never run out of good, clean water here."

Water ranks among the greatest worries of chile farmers across New Mexico, for adequate moisture is essential to successful chile growth. Too much of it, whether through excessive irrigation or rain, causes chile to wilt and rot. Too little of it, and chile starves, stunting pod growth or completely killing the plant. Fortunately, even on a bad snow year, when a poor spring runoff threatens farming across the state, Ranch O Casados doesn't lack for moisture. Even during the dry years, the water still flows, two-and-a-half-feet deep, into the Casados acequia.

Plentiful water was one of the two permanent blessings that would help Casados tend to his chile crop. The other was the birth of his only son, Orlando Anthony Casados, Jr., the youngest child in a family of six children. His father called him "Sonny" for short.

"I remember going to the hospital and this nurse comes out," Casados says. "I guess my wife had told her to say that it was another girl. I said, 'Gee whiz.' Then she showed Sonny to me and I said, 'Oh, wow!'"

From planting to harvest, Orlando, Jr., grew up learning the workings of a farm. When the freeze of '71 wiped out the apple orchard, young Orlando helped his father begin to build their chile crop. But later that year, Marie Casados died. Suddenly, the elder Casados lost all interest in the farm. "I decided to quit; I was just getting too old," he says. "I let Sonny take the farm. I figured the less I knew, the better I'd feel."

Still, Casados did not leave farming completely. During the next few years, as Orlando slowly boosted their chile production to fifteen acres of red and green, the elder Casados explored ways to market the crop. He came up with a plan to package their chile and other farm products for sale throughout the region and by mail. But in order to do so, his business needed a name. Setting up a pun on the word *rancho*, the Spanish word for ranch, Casados came up with Ranch O Casados. The letter "O" was for the two Orlandos who oversaw the farm.

"I was a little good at art, too," Casados recalls. "I went to Ranchos de Taos and drew a picture of the church with some hornos and the Taos Pueblo nearby. The idea was to show that Ranch O Casados produced chile the way the Spaniards and the Indians did here long ago."

Casados took his drawing to a press in Denver and had hundreds of packaging labels printed. He then began generating accounts for his products in New Mexico, Wyoming, Colorado, and Texas. He sold ground red chile, dried green chile, and literally thousands of double-stranded, six-foot-long red chile ristras for seventy-five cents a string. He personally delivered his products to his customers by truck until 1986, when he let his oldest daughter, Dolores, take over the packaging and the orders by mail and he began to enjoy the quiet retirement life. In 1992, he had his hip replaced.

"Now," he says, "I just listen to the birds sing in the morning and watch Sonny do all the work."

Orlando Casados, Jr., is standing next to a green Ford parked beside a twenty-two-acre field flanked by peach trees, cottonwoods, and an uninterrupted view of the Sangre de Cristo Mountains. Four acres of bushy "native" chile plants run lengthwise along the southern end of the field, and a wide acequia borders the field to the east. Purple morning glories and golden daisies punctuate the green growth like slender exclamation points.

The mid-September morning sun is warm enough for Orlando to be wearing shorts and a sleeveless shirt, but snow is predicted today in the Colorado Rockies. In preparation for that, Orlando has instructed two of his employees to spend the day hauling in as much of the remaining chile crop as they can. The men walk back and forth between the field and the truck, balancing bushels of green chile on their shoulders before dumping the pods into the bed.

"Once it gets cold up there in Colorado, it's only a matter of a couple of days before we get froze out down here," he says. "By the weekend, it'll probably all be gone."

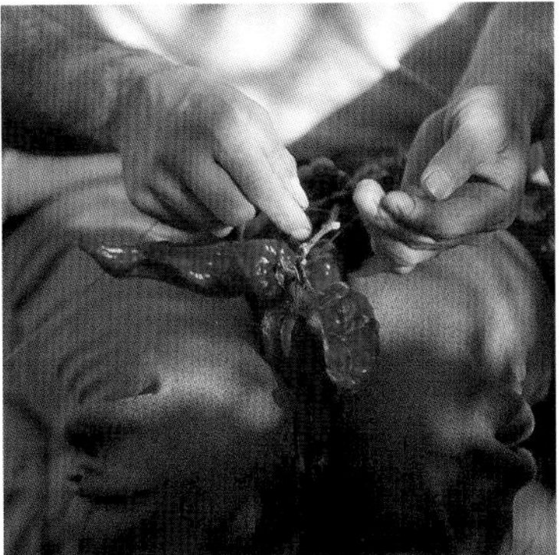

At harvesttime, Ranch O Casados workers string ripe red chile pods into ristras and hang them to dry.

At Ranch O Casados Farms, green chile is gathered into bushels, then slow-roasted over aspen wood in large adobe hornos.

And good riddance. Although these four acres have produced a good growth, it has been a tough year for chile at Ranch O Casados. "I guess it's probably one of the worst years I've ever had other than when we've been hailed out," Orlando says. He started out the 1995 season with twelve acres in all. He planted with confidence, for just one year earlier he had produced his best crop in ten years. That year, the weather was warm, there was not too much rain, and the cold weather didn't come until late October. After discouraging yields in previous years, Orlando had cut his acreage to five acres in 1994, but they turned out to be five of the most prolific acres he had ever grown. That success prompted him to nearly triple his acreage in 1995, but now, as the season draws to a close, he is left with six of the most difficult acres he has ever grown.

Ideally, Orlando would have planted the 1995 crop in early April, but a late spring cold spell kept him from planting until the middle of May. Hoping for a little extra luck, he planted on May 15, the feast day of San Isidro, the farmers' patron saint. He planted mostly native chile along with a little Española Improved and a bit of Big Jim. But the cold reappeared shortly after planting and lingered through the entire month of June. Because plants need to emerge quickly from the soil in order to establish a good stand and adequate yields, the low soil temperatures slowed the plants' initial growth. Discouraged, Casados replaced half of his chile crop with corn.

The remaining six acres finally started to set in the middle of July, about three weeks before Orlando normally would begin harvesting his crop. July and August, however, proved to be extremely favorable months as a record-breaking heat wave hit northern New Mexico. Chile is a heat-loving plant, and with temperatures in the Española Valley registering up to 105 degrees, Orlando's chile thrived. By the third week in August, the chile was mature enough to harvest, but about two weeks into the picking, in early September, cold settled into the valley again. Another chile season cut short at Ranch O Casados and at other northern New Mexico farms.

"With the exception of last year, it's been rough up here in the north the last five or six years," Orlando says. "The weather hasn't cooperated at all."

The twelve to fifteen acres of chile that he grows every year may sound small compared to the hundreds of acres grown by farmers in southern New Mexico. In light of difficult growing conditions in the area, though, Orlando's is one of the largest crops in northern New Mexico. From planting to harvest, the course of northern chile crops generally follows four to six weeks behind a southern crop. While southern farmers complain that too many insects are the result of too little cold to kill them off, northern farmers complain that too much cold cuts their chances of producing a successful crop virtually in half. Consistently moderate temperatures in the southern climes assure those farmers a minimum of a nine-month season. Northern farmers don't get much more than a five-month season, if they're lucky.

Farming is a risky business no matter where one undertakes the task, but the challenges in northern New Mexico place it among the least desirable ways to make a living there. Compared to the nearly thirty thousand acres of chile that are grown in southern New Mexico each year, only seven hundred acres are produced annually in the north, and those numbers are dropping fast. "There are less and less people planting anymore because it's so doggone much work and the weather's so bad," Orlando says. "A lot of people have quit altogether."

But those like Orlando who are committed to maintaining the agricultural traditions of the area have found ways to adapt. Many northern farmers swear by native, or local, chile cultivars because the plants have a higher tolerance for the extreme climate than do most commercial types. Although they are no less susceptible to problems caused by low soil temperature or frost, the local varieties produce early yields and large quantities of mature red fruit. Commercial varieties such as New Mexico 6-4 or Big Jim, produce large, smooth pods but tend to mature later and produce smaller crops. One exception is Española Improved, a hot, fast-maturing chile released at New Mexico State University in 1984. This type produces large, smooth, pungent pods and generates high yields of both red and green fruit.

"It's definitely a different chile we grow up here," Orlando says. "It's not as heavy or as thick-skinned as they grow down in the south, which makes a lot of difference in terms of weight and volume, but the flavor is totally unique."

By growing chile from seeds that are descended from ancient local chile plants, Orlando has attained a consistency in his product that is hard to match. And like his father before him, he relies on the one element that can put him at an advantage over other local farms: water.

In northern New Mexico, most chile acreage is irrigated using community-based acequias, or ditch irrigation systems, such as the El Guique Acequia. In the Española Valley, these systems depend on the Rio Grande as their main water source. Ditches in other parts of the region are fed by the Rio Chama, the Rio Embudo, and the Santa Cruz reservoir. Although the frequency of irrigation depends on plant size, soil type, humidity, and prevailing temperatures, most northern chile crops require an average of twenty-four to thirty acre-inches of water, applied in six to twelve irrigations, during the growing season. (An acre-inch is the amount of water needed to cover an acre to a depth of one inch.) That is considerably less than the four to five acre-feet demanded by southern chile crops. The need for less water is perhaps the only advantage of the shorter northern season.

A few hours later, Orlando is standing in a sea of freshly picked green chile that is ready for roasting. Two huge adobe hornos sit side by side under a shady steel roof, breathing slow puffs of hot air into the autumn breeze. Their fires have been stoked for three hours now; in each, slender logs of aspen wood smolder beneath metal racks that slide in and out of the fiery oven fronts. A giant metal trash can filled with water stands nearby. A man pours a bushel of chile into the water, then arranges the wet pods in a single layer upon the metal rack. He pushes the rack into the oven's dark, gaping mouth. The chile hisses as its wet skin steams, bubbles, and sears. When the pods are puffy, with their skins brownish black, the man retrieves the rack. He turns the pods over and then pushes them back into the horno to roast the other side.

Mariachi music from a dust-covered radio serenades the two chile cooks as they slide bushel after bushel into the fire. The roasted chile is dumped into plastic garbage bags and then packed in beer boxes for customers to carry away. Compared to propane-fueled roasters, horno roasting is a hot, tedious, and laborious task. An horno takes about twenty minutes to cook two bushels of chile. A gas roaster takes only five. "Roasters are faster, but they don't produce the same results," Orlando says. "Unless somebody really knows how to use a roaster, the roasting is rarely even. And if only half of a pod gets properly cooked, it is going to be tough to peel."

Orlando started horno roasting in the late 1980s after a hailstorm threatened his crop. The storm was quick and didn't destroy the chile, but it left the pods with damaged, pockmarked skin. Orlando remembered the old, unused horno that had been sitting outside his house since he was a boy. He fashioned a metal rack, just large enough to hold half a chile bushel, chopped some kindling, and fired up the oven. The chile came out tender and plump and the pockmarks could no longer be seen. Orlando placed ads in the newspaper and on the radio announcing that Ranch O Casados was roasting chile the "traditional" way. The ads lured customers from throughout the state, many of whom still travel hundreds of miles to Ranch O Casados for fresh horno-roasted chile every year. Orlando has since replaced the small horno with two much larger ovens, which he also uses to roast chicos from his homegrown sweet corn.

"It's the old-fashioned way of doing things, keeping close to the product. Every chile that goes into that horno gets handled by a person, and our customers appreciate that," Orlando says. "In the south, they're much more high-tech than we are up here. They have chile dryers and processors and laser levelers to get the rows in their fields straight. It's big business down there, but we're just little family farms. We're little and we choose to stay that way."

Indeed, there is a kind of nostalgia among northern farmers such as Orlando, a certain sentimentality for their small, hard-won crops. It makes one wonder

whether or not they would really want a longer, easier growing season if given the chance. Would their chile mean as much if the challenges were gone?

"I told my banker once that it would be a lot less trouble to go fishing every day than to try and grow a crop of chile. A package of bologna, a six-pack of beer, a loaf of bread, and that's it," Orlando says. "Sometimes, I get so disgusted and depressed that I feel like quitting, but come spring, you just feel like getting out there on that tractor and doing it all over again. Chile is our heritage. I just think that somewhere along the line, if we keep up the tradition, it will pay off."

Today, Ranch O Casados features twenty-two different chile products—spices, salsas, and fresh and dried green and red—that are sold off the farm and by mail to hundreds of regular customers throughout the United States. After purchasing other parcels throughout the years, the farm comprises about fifty acres in all, now worth about $25,000 an acre, compared to the $25,000 Orlando Casados, Sr., paid for his original farm. Though he is constantly approached by prospective buyers, the elder Casados vows he will never sell.

"Money's no good, money disappears, but a good chile crop lasts a lifetime in my mind," he says. "It's a wonderful thing, nature. One little seed will grow a plant and a plant gives you thousands of seeds in return. Down there in the south, a lot of those chiles are as big as a banana, but they taste like cardboard, no flavor at all. This is the best place for growing chile in the whole world."

Casados pivots himself around on his knobby cane and begins his short walk home, where he will take his customary afternoon nap. He walks past his son's gray stucco home and across the dirt road that leads to his pale pink adobe. Step, pause. Step, pause. Step. Slowly, he walks, talking every step of the way. "Farming is a nice little life, a healthy life," he says. "You never get rich, but it's a nice little life."

South out of Santa Fe, Interstate 25 will descend three thousand feet in elevation over the next 280 miles into the bountiful chile-growing country of the south. Like a reliable companion, the Rio Grande follows the same path, its fluent waters keeping rhythm with the ever-changing land.

A largely rural countryside prevails along the route, and the dominant piñon–juniper ecozone gradually yields to the drier juniper–grassland and shrub–desert that distinguish the low desert of the southern region from the high desert north of Santa Fe. Between Socorro and Truth or Consequences the parched sagebrush is interrupted by an ocean of blue. No, not an ocean

Elephant Butte Dam, the lifeline of southern New Mexico farmers. Bottom: At time of construction, 1916. Bottom photograph by Ernst Ruth, Sr. Museum of New Mexico Neg. No. 85777. Right: An aerial view of Elephant Butte Dam, ca. 1928–32. Museum of New Mexico Neg. No. 58276.

Butte-DAM-N.M.

Opposite and above: The Rio Grande is a plentiful source for the many irrigation ditches and fields that dot the landscape throughout southern New Mexico.

but a man-made reservoir at the southern end of which a gray, humpbacked mass of land rises from the water in the guise of a giant elephant. This is not a mirage, a specter of abundant moisture in the thirsty landscape, but Elephant Butte, a forty-thousand-acre, forty-mile-long reservoir and recreational lake and the irrigation lifeline of thousands of southern New Mexico farms.

Settlers first arrived in the region to take advantage of the Homestead Act of 1862, which allowed individuals to homestead 160 acres of public land, but repeated flooding of the untamed Rio Grande impeded most farming efforts. Plans to build a large dam on the Rio Grande for purposes of irrigation and flood control were initiated in 1896, but for years the effort was stalled by legal challenges from Texas, which feared it would lose water if portions of the river were contained. Finally, in 1911, construction of the dam commenced as a $42-million Federal Reclamation Project was jointly funded by the federal government and thousands of southern New Mexico farmers. When the dam, a graceful concrete construction of arches and spires, was completed by the U.S. Army Corps of Engineers in 1916, it was the largest in the world, and the 2.2 million acre-feet of water it contained made it the largest reservoir in the United States. The reservoir was called Elephant Butte because of the elephant-shaped island just upstream of the dam. Its creation completely transformed southern New Mexico from a dry, hot wasteland to a lush agricultural landscape where farming was central to the economy and culture of the residents who lived there.

Even before the butte's establishment, though, farmers further south in Doña Ana County, which encompasses the chile-rich Mesilla and Hatch valleys, had begun to develop an important agricultural niche in the region. Spanish colonists surely passed through the fertile Mesilla Valley in 1598 as they made their way north through Mexico to establish the first Spanish capital of New Mexico at San Gabriel, but the valley remained unsettled until 1850. The two-year Mexican–American War had come to an end in 1848 with the signing of the Treaty of Guadalupe Hidalgo, which ceded to the United States a sizable hunk of Mexican territory, stretching from southern California to modern New Mexico. A group of Mexicans not wishing to become Americans relocated from the east side of the Rio Grande to a small stretch of tableland on the west side of the river that was still part of Mexico. They called their new town La Mesilla, and in 1853 the Mexican government approved their petition for a land grant. However, less than one year later the Gadsden Purchase realigned the international boundary again, making La Mesilla part of the United States.

The settlement and its surrounding valley soon thrived as the population and agricultural hub of the southern New Mexico territory. In 1858, the town became a main stop on John Butterfield's popular Overland Mail stagecoach route that traveled between St. Louis and San Francisco. Waterman L. Ormsby, a correspondent for the *New York Herald* who was a passenger on the first Overland Mail westbound stage as it thundered through the Mesilla Valley, wrote of "irrigated fields groaning with the weight of heavy crops."

Nearly 140 years later, the Mesilla Valley holds the largest area of irrigated farmland in New Mexico. Extending south along the Rio Grande from Radium Springs to El Paso, Texas, the valley makes a diagonal slice through the center of Doña Ana County, an estimated one hundred thousand acres of land straddling both sides of the Rio Grande. The valley, cut into upper and lower regions with the city of Las Cruces as its approximate dividing line, is home to almost 1,200 farms, most specializing in the cultivation of chile, cotton, and pecans.

It was cotton that Andres Apodaca first planted after coming to La Mesilla from Texas in the early 1900s. In time he farmed chile to sell in surrounding towns, stringing ristras that his wife, Guadalupe, put on the clothesline to dry. The couple spent their lives in La Mesilla farming twenty acres and raising four sons. When they died, their sons inherited five acres apiece.

One son, José Apodaca, and his wife, Genevieve, worked those five acres until they could afford to buy more. In 1941, the couple moved to La Mesa, about ten miles south of La Mesilla, where they bought five hundred acres of undeveloped land. Dubbing the land "Apodaca Farms," José began building a sprawling white adobe hacienda with a red tile roof. By 1944, the house had grown to five thousand square feet, stretching two stories high and twenty rooms wide, and the land had given birth to fields of cotton, chile, alfalfa, and pecans. Like his father, José's main source of income was cotton. Chile was considered a lowly secondary crop.

Meanwhile, the Apodaca household came alive with two daughters, Mary Helen and Emma Jean. As soon as they were old enough, the girls went to work in the fields. "At that time, they still harvested cotton by hand," Emma Jean Cervantes now recalls. "To get us girls to work harder, they had us compete as to who could harvest the most. I always surpassed them all."

When the value of cotton began to decline in the mid-1950s, José put more acreage into chile. Soon, he secured his first major contract with the Old El Paso brand. Instead of cotton, Emma Jean and her sister now picked chile. "I was introduced to chile at a very young age and it became a part of my life," Emma Jean says, "but Dad was very traditional. He had two girls: One of us had to be a nurse, the other had to be a teacher."

And so it was. After high school, Emma Jean left La Mesa for Los Angeles, where she earned a bachelor's degree in nursing. Her sister went to school to become a teacher. In 1967, however, Emma Jean returned to La Mesa and persuaded her father into letting her work on the farm. During the next ten years, she helped him manage up to 250 acres of New Mexican green and red chile. In 1978, when José had a heart attack, Emma Jean was the natural person to take charge of the farm. She leased five hundred acres of land from her parents and then purchased four hundred acres more in La Mesa of her own. In 1981, she began to grow and process jalapeños and cayenne on a very small scale.

"I had about five years of my father's counsel before he died in 1983," Emma Jean says, "and right to the end, he was very traditional. He warned me that agriculture is a jungle."

Emma Jean Cervantes's grandmother, Guadalupe Apodaca, harvesting chile at the family farm, 1947.

In the U.S., more jalapeños are eaten on "nachos"—the salty, gooey, cheese-laden appetizer—than are eaten fresh, pickled, canned, or mixed with other ingredients in salsas. The hot, thick-fleshed jalapeño pod takes its name from its original home of Jalapa in the Mexican state of Veracruz, but it is estimated that the majority of jalapeño slices sprinkled on nachos today are grown in New Mexico. Along with New Mexican and cayenne chile, the jalapeño is one of the three major types of chile grown in the state.

In 1982, U.S. astronauts honored the jalapeño as the first chile to be taken into space. One year later, upon the death of her father, Emma Jean decided to make jalapeño and cayenne an even bigger focus of the farm than the traditional New Mexican red and green.

"We decided to really focus on different varieties of chile," she recalls. "I felt in my heart it was important to preserve what my father had worked so hard for and to expand on it."

Genevieve Apodaca, Emma Jean Cervantes's mother, in the family cotton fields, 1947. Below: Emma Jean Cervantes and her children, Kristina and Dino, have turned a small family farm into the largest chile growing and processing company in New Mexico.

A twenty-foot-tall red chile sculpture provides a colorful landmark for the Cervantes Enterprises building in Vado.

Within ten years, Emma Jean had expanded her father's small farming venture into Cervantes Enterprises, one of the largest chile-growing operations in New Mexico today, earning herself a nickname as "New Mexico's First Lady of Chile Production." In 1993, she planted 450 acres of chile at Apodaca Farms and subcontracted 750 acres more from other area farmers for a total of 1,200 acres in all. Spread among that farmland were nine varieties of chile, including mild, medium, and hot New Mexican varieties ranging from "Anaheim" and Sandia to Big Jim; four varieties of jalapeño; superhot habaneros; and enough acres of cayenne to earn her the distinction of being one of its biggest producers in the country.

It was also in 1993 that Cervantes Enterprises started shipping its products outside of the United States and into the international marketplace. For Emma Jean, the move confirmed her decision to expand her father's farm from a chile-growing business to a chile-processing business. In 1981, she began operating Cervantes Enterprises out of a plain yellow building in Vado, a largely Hispanic farming community in the central part of the lower Mesilla Valley. During its one-hundred-year history, Vado has survived at least four name changes and various incarnations. By the harvest of 1993, it had become home to the bustling business offices and large chile-processing complex of Cervantes Enterprises. The complex, which is tucked neatly behind Emma Jean's office and bordered by a tall chain-link fence, is a series of large metal buildings and open-air sheds. Across the street from the main entrance, in view of all who travel Vado's main street, is a twenty-foot-tall red chile wood sculpture that Emma Jean commissioned a local artist to carve in 1987. As statuesque and curvy as a pinup girl, the pod balances on its tip atop a stone foundation, its green stem pointing north.

"It's the only chile monument in the U.S.," Emma Jean boasts. "We grow chile, we process it, we market it, we distribute it, and we worship it."

Throughout the processing plant, ceilings and other surfaces are splashed with the plasma of ripened chile pods while pavements are littered with castaway seeds and skins. Gargantuan twelve-thousand-gallon plastic containers the size of oil drums are stacked throughout the facility labeled "Cayenne '91," "Jalapeño '91," and "Jala-Verde '92." Some are filled with thirty-five thousand pounds of pickled jalapeño peppers in a salt-and-vinegar brine. The chile, which has a storage life of up to three years, is warehoused and distributed from the facility year-round.

At 6 A.M. sharp, the facility comes to life as the first load of cayenne is trucked in from the fields and workers prepare to pulverize the pods into a pasty concentrate. An army of forklifts scrambles about the plant, conveyer belts begin to move, and uniformed workers don hard hats, rubber gloves, and cotton masks. To ensure the freshest product possible, all of the chile designated for processing will pass through the plant on the same day it is harvested. The workers' goal is to process some twenty thousand pounds an hour of cayenne.

The process begins as chiles are piled upon long conveyor belts and workers strip the pods of extra foliage and debris that have been carried in from the fields. When the chiles have been washed clean, the stems are plucked from the pod tops before they fall off the conveyor belt and back into crates. The acrid aroma of chile fills the air like an invisible fog. To the untrained nose, the effect of inhaling such huge concentrations of chile can be as caustic as inhaling pure ammonia: eyes water, throats itch, noses burn, sneezes erupt. Yet there is not even a sniffle among these workers, who appear to have become acclimated from their constant exposure to the pungent smell. From here, the cayenne peppers move to a machine where they are coated in salt or vinegar and crushed into concentrate. Finally, the mash is siphoned into the large containers before being shipped.

Emma Jean greets the workers as she strolls through the plant, bending every now and then to pick up a perfectly good pod that has fallen to the floor. In another room, a group of workers is destemming sun-dried red chile pods that will be packaged into powder or sold whole. The plant will operate at full capacity until midnight; in the peak harvest months of September and October, it will run twenty-four hours a day. Emma Jean knows how many pounds of chile pods pass through the plant every day. Good management, she says, is the only way to survive the agricultural jungle.

"Your water management, labor management, equipment management, financial management—all have to be very efficient or you can't survive as a chile farmer," she says. "Raising chile is like raising a child: It requires a lot of tender loving care. If there's an impairment in any part of the growth or development process, the production will suffer."

Overall chile production in New Mexico in 1995 decreased 20 percent from the previous year after a combination of bad weather and the deadly Curly Top virus devastated crops throughout the state. Farmers in southern New

At Cervantes Enterprises, chile seedlings are started in a greenhouse before being transplanted to the fields in March.

Mexico were among the hardest hit; in Doña Ana County alone, harvested chile fell from 8,200 the year before to 6,000 acres. At Cervantes Enterprises, however, 1995 has been a good year, so good that this Labor Day has been designated a day of work, not rest. Emma Jean, therefore, spends this holiday morning driving through the maze of dirt pathways connecting four hundred acres of chile spread over four Cervantes farms between Vado and La Mesa.

Most commercial chile acreage today is directly seeded, but some farmers, including Emma Jean, prefer the method of starting plants in the greenhouse and then transplanting them to the fields in March, a practice common in the industry before 1940. Adherents cite several advantages, such as a more efficient use of expensive seed, more reliable plant establishment and pod uniformity, the need for less irrigation and cultivation, and, perhaps most important, greater yields.

"Chile is one of the few agricultural crops that brings in good income, but its gotten more expensive to grow," she says. "Seedlings are insurance."

Emma Jean also has invested in a drip irrigation system on 180 acres of her land, a technique now heralded by growers and agronomists alike as the most efficient way to water chile. In addition to being a great water conservation practice in New Mexico's arid desert landscape, drip irrigation has been proven to reduce the incidence of chile disease while increasing overall yields.

"What we're finding is that our yields are double that of the traditionally irrigated crops," Emma Jean says. "It was a major financial invest-

Jesus ("Chuey") Maldonado, bottom left, has been the harvest crew supervisor at Cervantes Enterprises for twenty years. Guadalupe Sepulveda was the operation's farm foreman for thirty years before retiring in 1993. By mid-October, the red chile harvest is in full swing in the fields at Cervantes Enterprises.

A worker fills a bucket with red cayenne chile at Cervantes Enterprises, one of the top producers of cayenne in the United States.

ment, but we figure it will probably pay for itself within three years."

As she drives past thriving fields of New Mexico green, jalapeño, and cayenne, Emma Jean points out other fields that lie fallow and bare. Perhaps the biggest preventative measure she relies on is the regular rotation of her chile crops. "My dad used to say, 'The land will become like a soggy enchilada with too much chile on it,' and he was right," she says. "You can't get good yields of chile unless you give the earth a chance to renew itself once in a while."

Emma Jean's prudent farming practices contributed to the year's success, but another reason is that a chunk of her chile acreage was planted more than two hundred miles south of the border in Mexico. The decision to expand the operation into Mexico is part of Cervantes Enterprises' long-range plan to acquire enough land in enough different climates so that, one day, the company's chile will be grown, harvested, and processed year-round.

The 1993 passage by the U.S. Congress of the North American Free Trade Agreement (NAFTA), which opened up the U.S. borders for the import of Mexican products such as chile, as well as increasing problems with Mexican migrant labor in the New Mexico chile industry, has set many New Mexico farmers against anything to do with Mexico. Emma Jean, however, sees it as a business opportunity.

"NAFTA has made Mexico more competitive, but it's also made it easier for us to deal with Mexico than it might otherwise be. It took a while to develop a relationship of trust and commitment, but so far they've delivered." She pauses. "The border has caused some significant social problems in New Mexico, but on the other hand, the kind of labor we need to keep our operations going here is accessible."

Arriving in La Mesa now, Emma Jean drives along her father's namesake Apodaca Road, where a series of fields are numbered Apodaca 1 through Apodaca 12. Tractors are parked in the middle of every field ready to pull flatbeds stacked with field bins through the chile rows. Scattered about the thick foliage, their heads and shoulders barely visible above the long-limbed plants, are the workers who fill the field bins with chile. Cervantes Enterprises employs 200 to 250 workers per day at harvesttime, most of whom live in La Mesa or in the string of border towns nearby.

They have been picking since 6 A.M. Most wear white long-sleeved shirts with either straw cowboy hats or bandannas tucked under baseball caps, trying to keep as cool as possible in the morning heat. By 2 P.M., it will be too hot to continue picking, meaning that the wages for the day must be earned by that time. Workers sit, squat, kneel, or bend at the waist as they diligently travel from one end of the row to the other. Their weathered hands move swiftly through the plants, picking only the mature fruit and leaving the others to ripen on the vine. The action is tedious but lively as music plays and the workers talk, joke, and sing among themselves.

For each twenty-pound bucket of chile that he or she picks, a worker receives a plastic coin handed out by a supervisor after he or she has checked that the bucket is full of mature chile. Only then is the worker allowed to dump out the fruit. Each coin will be worth about $1.75 apiece at the end of the workday. Emma Jean estimates that one individual can pick an average of 1,000 to 1,200 New Mexico–type chile pods a day but less of the smaller jalapeño and cayenne pods. Most workers will pick an average of twenty to thirty buckets a day, bringing the company's harvesting cost to $8,000 a day. A recently installed computerized harvesting machine keeps track of the harvesting process. Now, when a worker comes to collect a coin for a bucket of fruit, he or she slides a card into the computer. The card identifies the worker and records the time and date that the bucket was dumped, monitoring each worker's progress and efficiency and telling Emma Jean approximately how long it takes a person to pick each bucket of chile.

"We like our operation to be proactive," she says. "We're trying to keep ahead of the industry in all ways."

Jesus ("Chuey") Maldonado, Sr., Emma Jean's crew supervisor for the past twenty years, and his family work year-round for Cervantes Enterprises, picking green and red chile from mid-July through December. January and February are spent sorting and cleaning the dried red chile, as well as starting the season's new seedlings in the greenhouse. In March and April, the fields are prepared and planted. From May to July, when the harvest begins again, the family helps thin, weed, and irrigate the new growth.

"*Es trabajo que me gusta mucho, tengo mucho confianza*," Chuey says. He wears his years of hard work in the deep, dark lines of his face but says he still enjoys the work nonetheless. Chuey's wife, daughter, and son are acting as field managers today, standing in the shade of an umbrella while inspecting the workers' buckets of chile and handing out coins. Like her husband, Chuey's wife, Ana, says she is satisfied with her work. "We have worked well and earned well," she says shyly in Spanish, but their twenty-

Cervantes Enterprises cayenne mash is siphoned into twelve-thousand-gallon containers and shipped to Louisiana, Arkansas, and elsewhere to be made into hot sauce.

Thousands of pounds of ruby red cayenne peppers await processing at Cervantes Enterprises. For every bucket of chile a worker picks, he or she earns a plastic coin that will be exchanged for cash at the end of the day.

Bins of ripe jalapeños fill the lot at the Cervantes processing plant. Above: A twenty-one-ton semi brims with mountains of whole cayenne peppers.

five-year-old daughter, Josie, who is fairly fluent in English, does not hesitate to say that she does not plan to spend her life in the chile fields. Her dream is to attend college in El Paso and become a social worker. Motioning to her fellow workers in the field, she says, "I want to help these people, do something for them and my dad and my mom, to make life easier for us."

Watching the activity from the opposite end of the field is Guadalupe Sepulveda, a bright smile and gray sideburns illuminating his face from beneath a black baseball cap. Emma Jean waves excitedly and then gets into her car and drives to the other side. When she arrives, the two give each other a long, warm embrace.

"*Eres mi angel de mi guardia,*" she tells Guadalupe, calling him her guardian angel. "This guy's known me since I was a little girl. He saw me grow up."

Guadalupe was sixteen with a work permit when he first came to the Mesilla Valley from Chihuahua, Mexico, in 1942. During the next twenty years, he earned his keep on various Mesilla Valley farms, eventually getting married, having six children, and becoming naturalized as an American citizen. In 1963, upon meeting José Apodaca, Guadalupe took a job as a heavy equipment operator at Apodaca Farms. In time, he was named farm foreman, giving him the responsibilities of overseeing the annual chile crops from planting through harvest.

"*Me gustaba mucho,*" Guadalupe says, explaining how the work and the chile became an important part of his life. He tells how every spring on the first day of planting he would kneel down in the dirt and pray that the seed would produce. Then he would wait for a month or so until the first sign that his prayers had been answered began to appear above the ground.

"Cuando viene el chile, cuando estan chiquitos, parecen helitas de oro," he continues. In the right light, just as the tiny chile plants were breaking through the soil, they looked to Guadalupe like little rows of gold.

After thirty years in the chile fields, Guadalupe retired in 1993. "It's stressful, but the stress just becomes part of your personality," Emma Jean says. "Like Guadalupe, my philosophy is I surrender a lot of what we do to God. You become very spiritual because so much of your life depends on the God-given: the sun, the moon, the wind, the water. I'm an early riser, and I like to get up and drive to the fields before anyone else is there and see what happened overnight. Tonight, I'll pray that everything we saw today will still be here tomorrow."

"What does your daddy do?" Emma Jean asks her six-year-old granddaughter, Alyssa. The child stands silent for a moment until a big smile spreads across her small face. "Pick chile," she says.

Emma Jean's children play key roles in the family business. Dino is the general supervisor and Kristina is the company manager. Eldest son Joseph is a Las Cruces attorney and a business partner in the company.

"My children grew up with farming—planting, irrigating, hoeing weeds. They feel very close to the earth," Emma Jean says. "A lot of young adults opt to have a job from eight to five, a secure paycheck. Farming is not like that; farming can be a twenty-four-hour, labor-intensive job. I feel very fortunate that my children chose it anyway. I know I will leave this company in very good hands."

Both Dino and Kristina left home for a while before deciding that home was where they wanted to be. Since their return to the business in 1989 and 1992, respectively, they have helped Emma Jean continue to take Cervantes Enterprises in new directions. The two know the chile business inside and out, discussing such things as market forecasting and economic impact as if commentators on the evening news. Together, they bring the perspective of a new generation to an age-old crop.

"We are trying to get away from that ma-and-pa type of operation, trying to become bigger," Dino says. "A lot of people are doing well with their sacks of chile on a street corner. We enjoy going out in the fields and dealing with the workers, jumping on a tractor, but we also realize that's not where our energies are best served."

"The whole industry is at a crossroads," Kristina adds. "From a sales standpoint, we think the market is just beginning to be tapped."

For Dino and Kristina, there is a whole world waiting to experience New Mexico chile. While many New Mexico chile producers look at the chile industry from the inside out, basing their production figures and marketing plans on what other New Mexico growers are doing, Dino and Kristina view the industry from the outside in. They are more concerned with what chile growers are doing in such places as Brazil, China, Mexico, or Europe than in Española or Hatch.

"My major competition is all overseas," Dino says. "I don't really have any domestic competition."

He explains that most chile is produced in tropical areas and grown as perennials. Although the southern New Mexico climate is far from tropical, he says there are few other places in the country that have as long a growing season to produce the high-quality chile that the international marketplace demands.

"Chile farmers in New Mexico are the best in the country. I've seen them in Florida, I've seen them in south Texas, I've seen them in California, and we are definitely the best," Dino says. "For that matter, we don't think that any of the world's growers can match us qualitywise, but because there are so many outside forces at work here, we can't match them pricewise. On the world market, growers can come in some 30 to 40 percent cheaper on chile mash and 15 percent cheaper on jalapeños than Cervantes Enterprises can afford to charge."

When Emma Jean's father started growing chile, he used to wash his seed with bleach and reuse it year after year. Now growers are required to buy "certified seed" approved by New Mexico State University. The regulated seed has helped farmers to produce more uniform crops with greater resistance to disease, but it also has added considerable costs to the crop. Since 1991, Dino says, the price of some certified seed has skyrocketed from $5 per pound to up to $1,000 per pound.

Still, the biggest cost of growing chile—an estimated 65 percent—involves labor. Cervantes Enterprises employs fifty-five full-time workers year-round between the processing plant and the farm, plus the extra 250 day laborers who work on the harvest crews. For each of those employees, the company pays social security and federal income taxes, as well as unemployment and workman's compensation benefits. The company also maintains monthly premiums on the $5 million insurance policy needed to cover the operation of its forty-seat employee bus. Each portable toilet that the company buys for every twenty workers, as is required by law, costs $2,500 more.

Forty thousand pounds of cayenne peppers are loaded for shipping, leaving a blanket of stray chiles on the ground.

"The people in the fields and in the plant are an important part of our business, and because we're pretty innovative, we seem to attract good, hardworking people that we're able to retain," Dino says. "I think that there's basically a good relationship between the grower and the laborer. I think it's that third power, like the government or some other entity, that has little knowledge about the industry that comes in and tries to shake things up."

Nonetheless, with the introduction of such things as drip irrigation and mechanical harvesters, Dino predicts that the need for a large-scale labor force will eventually be eliminated from the industry. The money saved can then be used to invest in more acreage, as well as in the equipment needed to maintain it. Plans for building new business offices and a new processing plant at Cervantes Enterprises are under way.

The ability of New Mexico chile growers to survive in the future also will depend on how much the state is willing to support the crop. "I think it's important for the state to start realizing how big a player chile is in our economy," Dino says. "You think of Wisconsin, you think cheese. You think of New Mexico, you think chile. Chile plays such a big part in the culture and history of New Mexico, and it's a big reason that the tourists come here. We all love it, but maybe we need to miss it for a while before we stop to think that we may be taking it for granted."

Emma Jean sits silently as she listens to her children talk about the business, responding only with an occasional smile or nod of her head. She is clearly confident about her children's expertise and their ability to guide Cervantes Enterprises into the future. As her business partners, Dino and Kristina have their mother's full professional respect, yet every once in a while Mom still steps in to guide them.

"I let my children have their own dreams, but I try to keep them grounded in reality as well," she says. "I've lived through loss, and I remind my children that one of these years we could completely lose all of our crops. I remind them to make sure they have the economics to survive the year when that happens. I remind them that we're not just growing chile or trying to make money, we're preserving our family heritage."

Owl Bar Café & Steakhouse

Rowena Baca, heiress to the green chile cheeseburger.

The sign that marks Exit 139 just south of Socorro is no different from those that direct drivers to other exits off of Interstate 25, but cars continually leave the highway there. Their destination: San Antonio, New Mexico. To many who travel between the northern and southern portion of the state, a stop at San Antonio's famous Owl Bar Café & Steakhouse is a must. Day and night, hungry customers converge upon the Owl, where the original New Mexico green chile cheeseburger was invented in 1948.

In 1945, when Frank Chavez returned to his tiny hometown of San Antonio after serving in the navy during World War II, he and his wife, Dee, opened the Owl Bar. The bar soon became the hangout for a handful of "prospectors" who had recently moved into town. In fact, these prospectors were the same atomic scientists who later activated the first atomic bomb at the nearby Trinity Site. They talked Frank into putting a grill in the bar and serving burgers along with beer. Frank added his own explosive ingredient to the all-American hamburger, heaping a pile of local green chile on top of the beef and cheese and creating the dish that has since become the Owl's claim to fame.

Today, the Owl is a dim, down-home kind of place run by

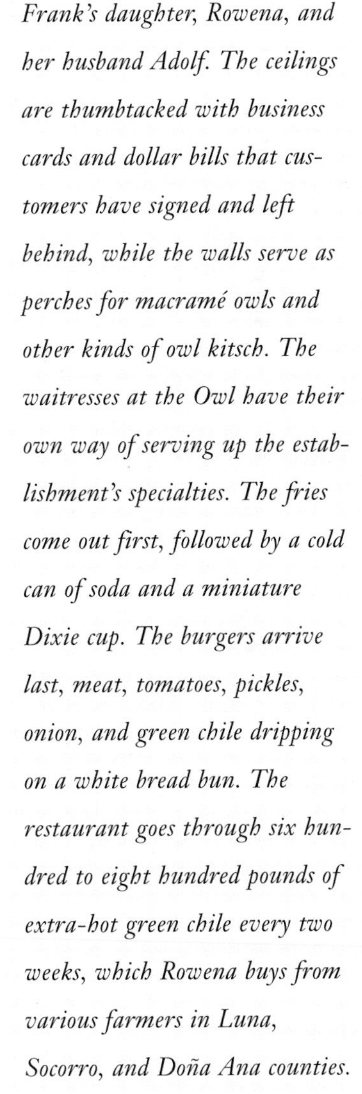

Frank's daughter, Rowena, and her husband Adolf. The ceilings are thumbtacked with business cards and dollar bills that customers have signed and left behind, while the walls serve as perches for macramé owls and other kinds of owl kitsch. The waitresses at the Owl have their own way of serving up the establishment's specialties. The fries come out first, followed by a cold can of soda and a miniature Dixie cup. The burgers arrive last, meat, tomatoes, pickles, onion, and green chile dripping on a white bread bun. The restaurant goes through six hundred to eight hundred pounds of extra-hot green chile every two weeks, which Rowena buys from various farmers in Luna, Socorro, and Doña Ana counties.

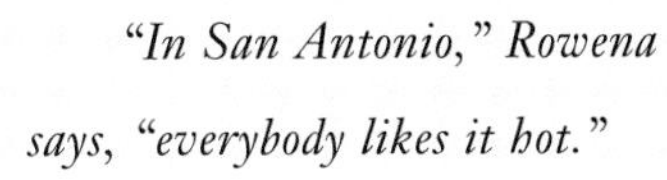

"In San Antonio," Rowena says, "everybody likes it hot."

Art and Agriculture

Cordelia Coronado, above, follows in the weaving and chile-growing traditions of her mother, Doña Agueda Martinez, right.

"The farmer's life is 50 percent luck, 50 percent hard work, and another 100 percent faith," Cordelia Coronado says. "Every year, I plant knowing full well that I may get nothing, but I have to try. Planting a seed is incredible. It keeps you close to the miracle of life."

Farming has long been a prized expression of traditional Hispanic life in northern New Mexico. In the small village of Medanales, where Cordelia's family has lived since 1925, farming goes hand in hand with another treasured Hispanic tradition: weaving. Here, chile farmers like Cordelia have been practicing their own distinctive textile traditions for as long as they have been planting chile. Weaving on massive looms with hand-spun wools, they work in what is known as the "Rio Grande" weaving style.

"My first inspiration is my mom," Cordelia says, "not only in the weaving but in the chile. Chile is my mother's cherished crop, and it's mine too."

It was Doña Agueda Martinez who led her daughter to a dual career in art and agriculture. Doña Agueda started working as a weaver soon after moving to Medanales in the mid-1920s to help supplement the family's farming income. She helped sustain the Martinez *rancho* by contracting her textiles to dealers in Chimayó, Española, and Santa Fe, eventually establishing a reputation throughout the region that brought collectors to her home. Doña Agueda divided her life between the rhythms of the land and the loom, farming from spring to fall, then weaving in the winter, when her farm was buried in snow.

"*Los rancheros aqui aprendieron conmigo*," Doña Agueda says, explaining that many northern New Mexicans sought her help in learning the secrets of growing chile at high-altitude terrains; others came to learn how to weave. Today at ninety-nine, Doña Agueda has established herself as the state's matriarch of the Rio Grande weaving style. With nearly seventy of her family members, spanning five generations, also practicing the art, she is head of the largest weaving clan in New Mexico.

Among Doña Agueda's most accomplished students is Cordelia, who like her mother is renowned for her weaving and her chile. Cordelia has spent the last thirty years in Medanales perfecting her own extra-plump chile pods, but during the winter, she creates textiles and teaches weaving workshops.

"I'm always working," Cordelia says. "When I'm not farming, I'm weaving. When I'm not weaving, I'm in the fields."

[Chapter Four]

Seasons in the Chile Cycle of Life

Standing side by side, Jim Lytle, Jr., and his son, Faron, survey the winter-worn field. The men cut the same stocky profile in the early morning landscape and carry the same playful look in their same-shaped eyes. When Jim lifts his right arm to keep the wind from blowing off his hat, Faron lifts his right arm to secure his baseball cap. There are twenty-two years and a lot of serious farming knowledge between them, but if one didn't know better, the two might be mistaken for a couple of country boys out to have a little fun.

It is 8:30 A.M. in late March in Salem, New Mexico. A small southern settlement five miles northwest of the Lytles' home in Hatch, Salem was established in 1908 by a group of New Englanders who named it for their Massachusetts hometown. This year, the Lytles have designated this crusty two-acre plot as the place to begin their annual chile-planting ritual. Altogether, the process will take place over three weeks and one hundred acres of farmland scattered throughout the winding Hatch Valley.

The season is chile. Chile season slips into the New Mexico calendar each year sometime between the first of March and the first of May. Enduring anywhere from five to nine months, it straddles the months that delineate spring from summer from fall; indeed, it outlasts them all. Chile is a warm-season crop, and much like tomatoes and squash, it requires long, frost-free days to produce good-quality, high-yielding plants. It prefers sixty-five- to ninety-degree temperatures and finds it hard to thrive at anything less. In southern New Mexico, thermometers usually mark the middle sixties by early March, prompting most southern farmers to get their planting under way. In northern New Mexico, temperatures often remain cold and chances of frost are high through late April or early May. Thus, northern chile crops will lag a month to six weeks behind their southern counterparts in terms of planting, growth, and harvesttime.

The sun is warm, the wind wheezy. Jim lifts the lid off a barrel of chile

Workers thin rows of young chile plants at a Lytle Farms field in Hatch.

Jim Lytle, Jr., left, and son Faron plant up to one hundred acres of chile per season throughout the Hatch Valley.

piquin seed that sits at the back of a green tractor, a combination planter and bed shaper that he and Faron built themselves. In general, a farmer can produce a good crop by planting two to three pounds of seed per acre, clumping three to five seeds six to ten inches apart. The Lytles, however, designed this planter and bed-shaper to compensate for potential plant loss from disease or other causes. It plants four rows at a time, shooting five pounds of seed per acre, twenty to twenty-eight seeds a foot, an inch deep into the dirt through a clear plastic tube. When the seed is released, a series of rotating discs throws a bed of dirt over the top, covering the seed and shaping the dirt into a neat mound.

Jim sticks a coffee can into the barrel, scoops out a canful of seed, pours it into the tube, and returns to the barrel for another scoop. Suddenly, the wind flings a seed into Jim's right eye, rendering him momentarily blind.

"Planting and harvesting are what I like the best," he says, struggling to clear his eye. "I like the challenge."

The seed expelled, Jim climbs back into the tractor and maneuvers its giant wheels between the wide furrows that separate the seven-hundred-foot planting rows. The field has not been planted with chile in three years; in an effort to prevent disease, it has been rotated with alfalfa, cotton, onions, and other crops. Although chile grows well on most New Mexico soils, a well-drained or sandy loam that holds moisture and contains some organic matter is ideal. The Lytles prepared for this planting by adding nitrogen and phosphorous, two essential chile nutrients, to the soil. The field was laser-leveled so there is no more than a one-tenth-foot difference from end to end. Such precision ensures that extra water drains from the field, reducing the risk of root disease. Then the earth was plowed, chiseled, disked, and smoothed into forty-inch-wide rows. Finally, about six weeks before planting, the field was irrigated to provide a moist, welcoming environment for the fledgling chile seed.

Faron mounts a wooden plank on the back of the tractor and catches his balance as Jim moves it slowly ahead. With deep, laborious groans, the tractor exhales a stream of cloudy exhaust into the cloudless sky. After two trips back and forth, the tractor stops. Jim and Faron leave the machine and trudge through the freshly formed rows to the center of the field and get down on their hands and knees. With fingers large and bare, they sift through the soil in search of seed.

"There's a lot of people that think you just go out and put a seed in the ground and it will grow," Faron says. He plucks a single seed from the dirt and holds it up to his father between his forefinger and thumb. "It's not that easy."

At this moment, Jim and Faron look like kids who have discovered an ancient treasure while playing in the sand. In some ways, that is not far from the truth, for Jim and Faron are but two in an extensive family tree of Hatch Valley farmers who have planted and sifted their way through acres of soil in hopes that something will grow.

"There are three languages spoken in this valley: English, Spanish, and Franzoy," Jim says. Spanish, he explains, is a testament to the centuries-old influence of Spain and Mexico in the region while English reveals the area's more recent American roots. "And Franzoy," he continues, referring to his maternal grandfather, "that's us."

Joseph Franzoi was born into a family of farmers in Tirol, Austria, near the Swiss border, in 1884. Celestina Formolo, his childhood playmate, later became his teenage love. In early 1905, twenty-one-year-old Joseph was to prepare for his induction into the Austrian military. Instead, he immigrated to America, leaving Celestina, his wife-to-be, behind. In Vulcan, Michigan, Joseph found work in the iron mines. He learned to speak English and began adapting to American ways. Finally, in an effort to somehow seem more American, he changed the "i" in Franzoi to "y."

Joseph and Celestina Franzoy immigrated to the United States from Austria in the early 1900s and became pioneers of the Hatch Valley chile industry.

Ten months after arriving, Joseph sent for Celestina, who made the eleven-day journey to New York aboard a German ship; she then traveled to Vulcan by train. They remained in Michigan for four and a half years, starting their own American family. Joseph envisioned a home amid the fabled wide open spaces of the American West. The family traveled first to Arizona and then New Mexico. After many years in Deming and working in the copper mines of Fierro, Joseph heard tales of another part of New Mexico that was rich in water and land. In 1918, he packed up his growing family again and traveled to a raw and rugged spot called Salem. It was located in a wooded area, or *bosque*, thick with dry scrub and ancient cottonwoods, on the banks of the Rio Grande.

Joseph, in the tradition of his European ancestors, was a farmer at heart. He purchased sixty acres of land and, with a team of horses, began to clear the tangled brush, an acre at a time, six acres in the first year. He planted beans, carrots, peas, potatoes, and beets and then hauled them in the back of a horse-drawn wagon to sell at nearby military forts, mining camps, and towns.

The Rio Grande still ran wild through the Hatch Valley at the time, and heavy rain often caused the river to overflow. Joseph had just established his crops when the heaviest rain the people of the valley had ever seen began to fall. For weeks and weeks, the water fell until many of the valley's newborn towns and farms had flooded. In Salem, the downpour was so fierce the combined force of the falling rain and the flowing river caused the Rio Grande to crest its banks and forge a new east-west course. The river cut straight through the center of Joseph and Celestina's farm, slicing the Franzoy homestead in half and forcing Joseph to cross the river to reach his fields.

Though New Mexico was by now officially part of the United States, many areas, including the Hatch Valley, were still largely Hispanic. The Franzoys were among the few Anglos living in the valley at the time. Not only were they still becoming accustomed to American culture, they were learning about New Mexican culture as well.

One day, one of their Hispanic neighbors invited the Franzoys to their home for dinner, where they were introduced to a vegetable they had never seen before. Their hosts called it "chile" and explained that it was a traditional part of the local cuisine. To Joseph, the strange dish didn't look very appetizing; its consistency was that of a thick European gruel, and it was red, the color of blood. Not wishing to insult his neighbors, Joseph took a taste.

Even before he had the chance to swallow, his eyes became teary and his mouth burned. Desperately, he turned to Celestina, lowered his voice, and whispered in his native tongue, "I think they're trying to poison us."

Two months have passed since Jim Lytle and his son, Faron, disseminated the season's chile seeds among their various chile fields, which are scattered throughout the valley to defray the risk of total crop loss to damage or disease. One field sits just beyond Faron's house in Hatch, situated next to a swath of fat, pearl-white onions, one of Faron's most prized crops. This chile field was planted two days after the Salem plot was sowed. Now, row upon row of chile plants are blooming, two inches tall with tiny clusters of leaves upon their fragile stands.

It is a crucial time in the chile cycle of life; active plant growth begins about two months after planting, blooming somewhere between mid-May and mid-June. Crowded plants produce much smaller pods than plants that have room to breathe and spread. When the plants are two to six inches tall, with two to six fully formed leaves, they must be thinned in order to maximize pod size. Ideally, single plants, or clumps of no more than three, must be uniformly spaced ten to twelve inches apart in a thirty-six to forty-inch bed, allowing potentially for a population of thirteen to fifteen thousand plants per acre.

***Opposite:** A field of newly tilled soil awaits the season's chile seed. **Above:** Two to three months after planting, chile plants sprout flowers and must be thinned to ensure maximum pod size.*

There is an art to thinning chile, depending on the type of chile being grown. This seven-acre field, spread with a thick blanket of Sandia chile, will be sculpted into shape by workers whose only tools are a hoe and an ability to determine which plants should stay and which should go. This morning, Faron has assembled a small work crew, two women and four men, to undertake the delicate task. They arrive eager to begin the day's work, for which they will be paid the standard hourly minimum wage. Faron, who grew up working in his father's chile fields alongside Mexican migrant workers such as these, greets them like old friends, speaking a Spanish as fluent and familiar as their own. He tells them they will be thinning Sandia chile today, and they head into the field, for that is all they need to know. Unlike the bushy Big Jim chile plant, named for Jim's father, they know that Sandia grows taller and requires less space between plants. While Big Jim is thinned to two plants every twelve inches, the formula for Sandia

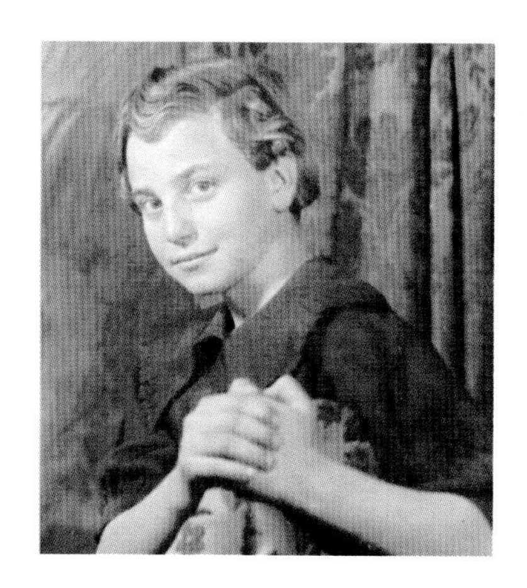

Top: June Franzoy, shown here at age twelve, grew up working on her father's chile farm. Bottom: Big Jim and Little Jim Lytle, farmers at heart.

is two plants for every nine.

"It don't pay to be mean to the people," Faron says as he watches the crew begin to work. "I need them as much as they need me. Like everybody, we've had our share of problems with labor, but I couldn't do this on my own."

On average, one person can complete one acre in the course of a seven-hour workday, but this particular crew works faster than most, giving Faron the confidence that they will complete all seven acres today. Working side by side, hunchbacked over long-handled hoes, they move quickly along the seven-hundred-foot rows, severing the roots of the unnecessary plants with steady, meticulous chops. Not even the pretty pink-and-purple morning glories that dot the field are safe. Technically, the trumpet-shaped flowers are weeds, and weeds compete with chile for sunlight, nutrients, water, and space.

The task is monotonous and tiresome, and yet it is only the beginning. Within the next six weeks, the field will be irrigated and constantly monitored for infestation by damaging pests, diseases, and a host of other unwieldy weeds. About mid-June, the plants will sprout a single flower at the first branching node. After that, the number of flowers will double at each new branch node. If daytime temperatures remain between sixty-five and ninety-five degrees, and if nighttime temperatures do not exceed seventy-five degrees, the fruit will start to set. Usually, the chiles will grow to full pod length within four to five weeks, normally reaching the mature green stage thirty-five to fifty days after the first flower blooms. Only then can the green chile harvest begin, about early August in southern New Mexico and later in the month for the northern crop.

"Our ultimate goal is to have every plant looking exactly the same," Faron says. He picks a pretty pink morning glory from the dirt and strokes its deceptively harmless bloom. "But nature works in mysterious ways."

The year was 1924. A few years had passed since Joseph Franzoy had taken his first taste of chile. He and Celestina had long since acquired a taste for the local favorite and, like many farmers in the Hatch Valley, had begun to grow New Mexico No. 9, the improved chile introduced to area farmers by Fabian Garcia. The agricultural potential of the Hatch Valley, by this time teeming with farms, was being realized. Joseph's family farm had slowly expanded to include cotton, wheat, corn, and hay, plus a small dairy and orchards filled with peaches, apples, and pecans. His horse-drawn wagon had been upgraded to a Model T Ford truck, enabling him to transport greater quantities of produce to markets in Deming, Silver City, and Fort

Baird. To this diversity, Joseph was preparing to add ten acres of chile. If it grew, he would sell it for ten cents a pound.

The Franzoy children now numbered ten, the last child a girl named June. Like the rest of her siblings, June grew up helping her parents on the farm. Of all her father's crops, chile was June's favorite. Unlike potatoes or hay, which her father's family had been growing for generations, chile, with its peculiar customs of cultivation, was as foreign to her father as he to his New Mexican neighbors. Every crop—indeed, every pod—had a personality of its own. Figuring out just what made the chile mild or hot intrigued Joseph, and he strived to make every plant reach its fullest potential heat, color, and size. Red chile or green, Joseph saw chile as one of farming's greatest challenges. June, however, saw chile through a child's eyes; it was the crop that brought the greatest look of satisfaction to her father's face. Growing chile somehow made her father feel like he was truly part of his adopted home.

By the time she was ten, June's father had mastered many of the challenges of growing chile, although cotton remained his main cash crop. Believing that there was a future in chile, Joseph had slowly increased his acreage to twenty, thirty, fifty acres a year, negotiating contracts with such companies as Ortega Foods. Every fall, fresh New Mexico green chile was shipped from the Franzoy fields to California, Louisiana, and other processing locales. By 1940, Joseph was one of the area's major chile producers. In the Hatch Valley, "Franzoy" was fast becoming one of the most respected names in agriculture. By then, June Franzoy, a high schooler, had met and fallen in love with Jim Lytle, a runaway from Hereford, Texas. They married in 1942.

Some fifty-plus years later, June Franzoy Lytle Rutherford is standing at the kitchen counter in her spacious, suburban-style home, mixing iced tea and talking about how she doesn't have time to talk. All around, shelves hold the trophies that June has won throughout the years at the Hatch Chile Festival cooking contest for favorite family recipes such as smoked turkey chiles rellenos, chile salad, and green chile and Spam®.

"I've always been on the farm," she says in a fast, scratchy twang. "This kind of work don't leave much time for talking."

June lives in Salem at Junction 390 in a house ringed by stately globe willows. The house sits on the same thirty-five-acre plot that Joseph Franzoy gave to June and Jim as a wedding gift, along with a cow, a calf, and a dozen chickens. Shortly after, their first child and only son, Jim, Jr., was born. In an effort not to confuse father with son, they called Dad "Big Jim" and the boy "Little Jim." Two more children, two girls, soon followed. So did a field full of chile.

June Franzoy Lytle Rutherford stands beside the stone grotto she had built in the backyard of her Salem home. Dedicated to the Virgin Mary, the shrine also hosts a statue of* San Isidro, *the patron saint of agriculture. "You have to have a lot of faith to be a farmer," June says.

"My husband didn't even know what a chile was, never even saw one, until he come here," June says, "but I knew Jimmy was going to be a good farmer cause chile was in his heart."

June and Jim started out small, planting eight to ten acres of chile a year. With every crop, important lessons were learned. Chile seeds will not germinate at low soil temperatures. Too much salt, either in the soil or in the irrigation water, burns and can eventually kill a chile plant. Too much wind causes young seedlings to dry out, and more mature plants to snap off at ground level. Phytophthora Root Rot disease, more commonly known as "Chile Wilt," causes plants to wilt and die. The disease develops when water sits in a field for long periods at a time because of overirrigation, heavy rains, or both. While some factors were controllable, others, like too much wind or rain or cold, were left to nature alone.

Growing chile was an ongoing process of trial and error; some crops survived, others died. Jim hauled his more productive harvests north to sell in Santa Fe and Taos, but the very best plants, those with the most uniform pods, the Lytles saved for themselves. After the pods were picked, their seeds were carefully removed. The seeds from the finest fruit were planted the following year.

"It takes about six or seven years to perfect a chile. You have to plant quite a bit before you get what you want," June says. "You plant it and then you select the best plants. You save the seed and plant it again. You just work with it until you get what you want."

Forty years earlier, in his seminal paper published on New Mexico chile in 1908, Fabian Garcia had written: "New Mexico chile growers do not pay any attention to the selection of the seed, and as a result of this, we are producing a very variable product." Garcia had since worked with area farmers to develop the highest quality seed possible.

By 1955, the Lytles' meticulous attention to their chile had caught the attention of Roy Harper, who had replaced Garcia as the chile researcher at New Mexico State University. The father of the mild New Mexico No. 6 chile, Harper was in the process of developing a new hot pod. He called on the Lytles to see if they would be interested in sowing his Sandia seed, providing the foundation seed, which had been tested in the school's experimental chile garden. He showed them an example of a perfect Sandia plant that had been grown in the test plot and challenged them to grow the same perfect plant in the field. If the Lytles could grow high-quality plants that produced pods that were consistent in size, shape, and degree of heat, the seed would be certified and sold for widespread use.

Every sack of chile sold at Lytle Farms is sealed with a "Big Jim Produce" tag in honor of Jim Lytle, Sr., who started the family farm in the 1940s.

"Roy gave us five or six ounces. We planted them and then saved the seed off the plants," June says. "Pretty soon we had five or six acres. The university liked the way we did it. They started releasing more seed to us."

After Sandia, which the university released as certified seed in 1956, the Lytles grew the medium-hot New Mexico 6-4 that Harper had been developing since the early 1950s. Certified seed for that cultivar was finally released in 1958 by Harper's successor, Roy Nakayama. The Lytles continued to work closely with Nakayama in growing the university's experimental seed. Eventually, the couple became certified seed dealers themselves, among the few authorized by the university to sell the approved seed varieties. Producing eight to ten varieties of certified chile seed at a time, the Lytles soon became two of the major seed dealers in the United States.

New Mexico's fledgling chile industry was growing fast; with more varieties of seed approved and vast improvements in the quality of the crop, more farmers began devoting land to the plant. The growth piqued Jim's interest in other facets of the industry, particularly the need for more efficient processing techniques. Before red chile could be ground into powder, for instance, it had to be dried. At the time, farmers placed their fresh-picked chile on rooftops and hillsides to sun-dry; every year at harvesttime, the Hatch Valley wore a coat of green and red. The technique may have worked for generations of farmers, but June didn't like putting her shiny, smooth, clean chile on the ground to dry. Birds and rodents contaminated it. It got dirty.

"We had an acquaintance, a little old man from Arizona, who came over here to see us, and he saw all the chile being dried on the hillsides," June recalls. "And he said to Jim, 'Why don't you just build yourself a dehydrator?' My husband said, 'Are you kidding? I don't have no money.' And he said, 'You don't need no money. I've got everything you need.'"

"So in 1959, we went over to Arizona. That man had barns full of equipment! Jim and I spent about a week over there. We took our tape measure and climbed all over that equipment. We measured height and width and length. We figured how much static air pressure we needed to circulate hot air through the drying tunnels. Then we came home and began to build one ourselves."

Jim Lytle, Jr., proudly displays red chile ristras made from chiles grown at Lytle Farms.

The moisture content of a ripe red chile pod ranges between 65 and 80 percent. For drying purposes, the moisture must be reduced to 8 to 12 percent. A dehydration machine would need to reach a temperature between 140 and 150 degrees to properly dehydrate the fruit.

"Everybody told us we was crazy, we was stupid, we didn't know what we was doing, we thought we were millionaires, we was throwing money away," June continues. "We weren't sure what we was doing either, but we was gullible enough to do it.

"So we got it all built and got ready to dry our chile. And everybody and their dog came to see the stupid idiots who didn't know what the hell they was doing." June sips her tea and laughs a long, satisfying laugh. "And it worked nice. It worked real nice. Now there're dehydrators everywhere."

"The dehydrator changed the industry enormously," Jim, Jr., confirms. "We went from drying eight to ten tons to fifty to sixty tons. Dad and Mom were industry pioneers. What I learned, I learned from them."

Jim is standing on the rim of a cement irrigation ditch that runs along the north end of a chile field about six miles west of downtown Hatch. It is late June, midafternoon, and he has come to check the progress of his crop. The crop is half New Mexico No. 6-4, a medium-hot chile, and half "*Lumbre*," an extra-hot pod that, Jim jokes, "is so doggone hot, it should be labeled a controlled substance." He leaps across the ditch and into the field to examine the young pods and is pleased with what he sees—the fruit has started to set.

"Ninety percent of your chile has to do with the soil," Jim says. "Chile

Chile Diseases

The *Curly Top* virus, the state's most destructive chile disease, devastates chile crops by attacking plant leaves and roots and stunting growth.

Weather plays a critical role in agricultural success or failure, and then there is disease.

ALFALFA MOSAIC VIRUS

As the name implies, this virus is found in alfalfa fields. The virus causes mild stunting, whitish, blotchy leaves, and, in some cases, distorted fruit. Because the insects carry the virus from infected alfalfa plants to chile, the virus commonly appears when chile is planted next to an alfalfa field or in a field where alfalfa had been grown the previous year.

BACTERIAL LEAF SPOT, CERCOSPORA LEAF SPOT, AND POWDERY MILDEW

These are leaf diseases whose symptoms frequently involve spotting, discoloration, and defoliation.

BLOSSOM END ROT

The lower half of a chile pod develops soft, black, watery lesions that are associated with inconsistent watering, calcium deficiency, and heavy application of nitrogen fertilizers.

(continued on page 80)

likes to have air, and it likes to be watered in a certain way."

The soil is still moist from the last irrigation five days ago, and it sticks to the soles of Jim's work boots as he continues down the row. Chile, a shallow-rooted crop, needs four to five acre-feet of water between planting and harvest. In general, however, frequency of irrigation depends on the amount of plant foliage, as well as on fluctuations in temperature, sunlight, humidity, and wind.

It can take up to three irrigations to establish a crop. Once small plants have emerged, irrigation may not be necessary for three weeks or more, but come late June and early July, fields should be irrigated every five to seven days. When the summer rains come, that interval is expanded to every seven days or more, depending on the amount of rainfall.

The leaves of a chile plant are a good gauge for determining how often to irrigate. In hot, dry weather, plants may wilt in the late afternoon, as early as the first day after irrigation. As the soil dries, the wilting may appear earlier in the day, signaling that irrigation is due. Decreasing irrigation frequency at the end of the season promotes ripening and improves the color of the fruit. In addition, a way to drain the field is necessary throughout the season. If water stands in a field for more than twelve hours at a time, Phytophthora Root Rot disease may result.

Jim spots Juan Contreras, his irrigator for ten years, standing by the ditch. In Spanish, Jim instructs Contreras to water only every other row. That way, the water will push any accumulated salt to the inside of the dry row, where it can't burn the plants. Contreras gets in his car and drives away. He will be back to irrigate in two days.

"Juan is like family," Jim says, watching him go. "The Mexicans, like Juan, who have been here a long time, we all get along. It's the problems with the new labor that has so many farmers snowed under."

When eighty-five-year-old Joseph Franzoy died in Salem in 1969, he left a huge extended family of farmers. Farms bearing the Franzoy name dotted the landscape from Salem to Las Cruces. He died a Hatch Valley legend, for partly through his efforts, an agricultural legacy—chile—was spawned. That same year, June and Jim Lytle began work with Roy Nakayama on a new chile breed. Nakayama sought to develop a bigger, thicker chile suitable for canning and cooking, one even bigger than the popular New Mexico 6-4, which grew six to seven inches long. Nakayama crossed pollen from a tiny Peruvian chile with other New Mexican varieties and labeled the experimental chile "No. 1084." After getting a good start on the plants in the university's experimental garden, he gave the Lytles some seed to grow. And did it grow: nine to ten inches long, with shiny skin and flesh as fat as a thick orange rind.

"We went crazy when we saw this big ol' pretty chile," June recalls. "You know farmers, they like everything big and pretty."

Like all new breeds of chile, though, this, too, would take years to perfect. Jim took a particular interest in the breed and worked closely with Nakayama in picking the seed to grow the following season. Besides the chile's canning potential, Jim believed that No. 1084 could be the ultimate chile for making chiles rellenos. In the spring of 1970, the Lytles planted the No. 1084 along with their other chile crops. Jim hadn't been feeling well, but he committed much time and energy to watching over Nakayama's wondrous new breed. He doted over No. 1084 as if it was the last crop he would ever grow.

In fact, it was. Jim had cancer. As his strength and energy waned, he occupied his mind with the progress of the year's chile crops, particularly the No. 1084. He saw the crops through from planting, thinning, and irrigating to the first flowers and fruit set. He watched as the plants stretched hip-high and grew bushy and full. He saw the pods grow long and firm in the dry summer heat, and as the summer wore on, he watched the green chile skins take on their ruby hue. As the life cycle of another chile crop came full circle, Jim grew sicker.

The harvest got under way. Every day from 6:30 A.M. to 1 P.M., laborers filled the Lytles' fields and carefully picked every chile by hand. But on October 10, 1970, the workers fell silent. Despite a brief, hopeful turn for the better, Jim Lytle was dead. Like his father-in-law, Jim died a Hatch Valley legend. He died trying to think of a way to build a mechanical chile-picking machine.

Jim would never see a full-grown No. 1084 chile pod, but for the next four years, with the memory of her husband still fresh on the earth's surface, June planted the seed. Each year, as the pods grew bigger, so did outside interest in Nakayama's new chile. "Everybody wanted it," June says. "Everybody in California tried to get me to sell it to them. They offered me all kinds of money, but I said, 'No. It belongs to Roy.' I don't do business that way." In 1974, June's No. 1084 pods grew perfectly. "I had this crop out here and it was beautiful," she says. "Nine acres of the prettiest chile you've ever seen." Then suddenly, on October 22, hail came. Surely, June thought, the crop is lost. "The hail was that deep," she says, her right hand building an invisible hail pile beside her. "It was so cold, but we got in there in the mud, and we picked that chile, and we saved the seed." The next day, Nakayama came to inspect the hail-pecked pods. Despite the surface damage, he deemed that No. 1084 was ready to be released as certified seed. "What should we name it?" Nakayama asked June. "I don't know," she replied. "It's your chile." When Nakayama returned to the farm a few days later, he carried with him one of the blue certification tags with which the university labels its certified seed. The tag read: "NuMex Big Jim." "Jim's dead," Nakayama said to June, "but his memory will live forever."

June's mother, Celestina, died in 1976. A few crops later, June's giant Big Jim chiles earned her a distinguished place in the *Guinness Book of World Records* for growing the biggest chile ever recorded. The pod was 13.5 inches long, though she says she has since grown a 17.5-inch Big Jim. "Our knowledge is our experience," June says. "We're not agronomists or nothing like that. We're just farmers."

Jim Lytle, Jr., was twenty-seven years old with two sons and a farm of his own when his father died. Soon after, he bought another 170 acres of farmland with his mother in Rincon, an old railroad town five miles east of Hatch. The way he saw it, more land and more work would leave his mother less time to think about her loss.

"I guess one reason I stayed here in Hatch was that when Dad died, I had to take his place," Jim says. "I had to take care of my mother and my sisters. I had to take care of the chile."

Taking care of chile is a major task. Seeing a crop through from planting to harvest requires extensive knowledge of the various things, such as diseases and certain environmental conditions, that can threaten a plant's survival. At the time Jim took over his father's chile farms in the early 1970s, disease control was becoming much more of a consideration in the maintenance of a chile crop. Chile acreage in New Mexico increased more than sixfold between 1960 and 1970, and increased acreage in any crop translates to an increased risk of disease. Disease occurs when some external factor disrupts the normal growth and development of the plant. Parasitic diseases are spread from one plant to another by infectious disease agents, such as insects, while nonparasitic diseases can be caused by certain nutrients, chemicals, or even the environment itself. Disease profoundly affects both

CUCUMBER MOSAIC VIRUS
This virus is carried by aphids who visit urban areas and feast on infected perennial plants and then transfer the disease to rural chile fields. Diseased chile plants exhibit elongated leaves with yellow-and-white spotting.

PEPPER MOTTLE VIRUS
A disease that is transmitted by aphids and is found in chile fields throughout New Mexico every year. Leaves become misshapened and puckered with light and dark patches, and the fruit grows smaller than normal. The disease usually makes an appearance in chile fields in late summer or early fall, reducing red chile yields.

PEPPER WEEVIL
A particularly destructive pest that feeds on chile pods and leaf buds and leaves its eggs and larvae behind to grow and feed on flowers, buds, and fruit. A kin to the ferocious cotton boll weevil, the pepper weevil causes premature fruit to drop from the plants; in a thoroughly infested field, it can result in up to a 50 percent crop loss. Sometimes entire fields must be plowed under because there are too few fruit left to harvest and the infestation can pose a threat to later plantings of chile.

PHYTOPHTHORA POD ROT, BLACK MOLD, ANTHRACNOSE, AND BACTERIAL SOFT ROT
These are diseases that cause rotting, lesions, and deformity in chile pods.

PHYTOPHTHORA ROOT ROT (A.K.A. "CHILE WILT")
Causes the plants to wilt, turn straw-colored, and die due to excessively wet soil.

RHIZOCTONIA ROOT ROT
Occurs in the seedling stage of a chile plant and attacks plants on the lower stem near the soil, causing roots to rot as the fungus moves up and down the stem.

ROOT-KNOT NEMATODE
This is not a fungus, it is a microscopic worm called a "nematode" that takes up residence in the roots of a chile plant. As a result, the roots develop small knots that prevent water and other nutrients from reaching the plant.

(continued on page 82)

the quality and quantity of a crop. For a farmer, it can make the difference between profit and loss.

In 1972 Jim faced the drastic consequences of disease for the first time. It was a bad year for beet leafhoppers, the insect that spreads the Curly Top virus, the most destructive chile virus. Severely stunting plant growth and killing off its roots, the virus wiped out the Lytles' entire chile crop. Nonetheless, the following year, the Lytles planted even more.

"Farmers see the world through rose-colored glasses," Jim explains. "You don't know whether the bugs are going to be bad, the market's going to be bad, or what, but you take the gamble."

The gamble involves various other viruses that can have detrimental effects on a crop. But not all chile pathogens are the result of insects or other disease-bearing agents. In just seconds, high winds or severe hail can cause extensive damage to even the healthiest chile plants. Excessive salt in soil or irrigation water, as well as excessive application of fertilizers, can also injure plants. And just as humans can suffer from sunburn, chile burns when shaded pods are suddenly exposed to intense direct sunlight. To minimize such risks, growers learn to choose their chile cultivars carefully, growing disease-resistant varieties that are well-adapted to local conditions.

Still, neither years of experience nor the most meticulous maintenance program guarantee a successful chile crop. For farmers such as Jim, dealing with the uncertainty and risk that accompany every stage of a plant's growth makes growing chile as stressful as any job on Wall Street. In an effort to ease the stress, Jim spent many years coaching the local Little League. Hanging around with the kids made Jim feel young, and Jim's son, Faron, the team pitcher, was a natural.

By the time Faron was a junior at Hatch Valley High School, he was throwing a ninety-five-mile-an-hour fastball. His senior year he earned a baseball scholarship to Western New Mexico University in Silver City and left Hatch to concentrate on his education and his pitching. In 1984, as a sophomore in college, he was drafted by the St. Louis Cardinals for their farm team. But his career in the majors would never come to pass. That year, a pitching injury forced him back to Hatch, where he married his childhood sweetheart, Rosie, and settled into a life at home on the farm.

Three years later, at age forty-four, his father Jim succumbed to the stress of farming and had a stroke. Faron's older brother had long since left Hatch to work in Las Vegas, Nevada, so it was Faron, who had grown up at

Picking red at Lytle Farms.

This field of chile, as well as many others in southern New Mexico, fell victim to the dreaded Curly Top virus in 1995.

TOBACCO MOSAIC VIRUS
One of the most common and potentially severe of all chile viruses is transmitted by mechanical means—human hands, cutting tools, and other farming equipment. Disease symptoms include foliage with raised bumps and mottled light and dark areas of green. Pods affected by the virus ripen unevenly and are reduced in size.

TOMATO SPOTTED WILT VIRUS
Small insects called thrips spread this virus to healthy chile plants from diseased host plants that range from tomatoes to peanuts to pineapple. The virus winters in various perennial weeds, such as possession vine or curly dock, that commonly grow near chile fields and affects chile pods in their later stages of maturity. Fruit in both green and red stages can be infected. On red chile pods, the disease is displayed by patches of yellow that never turn red while green pods develop small, off-colored spots.

VERTICILLIUM WILT
Causes wilting when fungus penetrates and clogs a plant's roots and stem.

his father's side learning, and loving, the challenges of growing chile, who assumed the major decisions and operations of both his father's and his grandmother's farms. It took Jim a year to recover. Then, just as he was ready to get back to work in 1989, he had a heart attack. Faron still loved pitching, but with Jim's poor health, he forgot about baseball forever.

"Faron was a heck of a pitcher. My dream was for him to be in the big leagues," Jim says. "I still think he could have made it if he had stuck with it."

Faron Lytle is standing in a six-acre field across the street from his grandmother's house, knee-deep in Big Jim chile plants. His father is a few rows away on bended knees, examining the chiles that bear his own father's name. "Everything I know about farming," Faron says, "I learned from my dad."

It is the first week in August, and the heat is sweltering. The dark green pods under Jim's inspection are extra-large with bluntly pointed ends. They are smooth, fat, and firm to the touch. They are ready for picking. The chile, which will continue to bear fruit until it is killed by frost, will be harvested now to promote further growth. If all goes well, in four to five weeks the plants will be harvested a second time. If the season is particularly favorable, a third set of fruit will develop and mature.

Meanwhile, other chile crops will be harvested only once in the green stage or, in some cases, won't be harvested at all. Instead, the chile will be left to turn red and sold as whole pods or will be dehydrated and ground into flakes or powder.

Every harvest is the culmination of years of planning, yet no matter how specific the plans, there are never guarantees. "Farming is a year-round process, and it's not just the plowing and the land preparations," Faron says. "Every year, we go to our banks, talk to our bankers, pay off our debts from the last year, and start to plan for the next. Usually, we make a five-year plan: where, what, how much we're going to plant. We try to stick to the plan, but sometimes the plan changes because of the market, disease, or some other unexpected thing."

Jim is standing now at Faron's side, laughing as he recalls all the times his farming plans have gone awry. If he had to choose the worst time, it would have to be the summer of 1992—and he wasn't laughing then.

Three hundred acres of chile were sitting in one of Jim's Hatch fields, plump, pretty, and ready to be picked. It was late afternoon when raindrops began falling from the sky. Then they started to freeze in midair. Soon large hail pellets were falling in force, landing with a thud, then a bounce, as if someone were throwing rubber balls from above. As they made their way to earth, the hail balls nicked and cut and pricked the foliage and the fruit of Jim's chile plants.

When the hail stopped, Jim drove to the field. The storm had lasted less than ten minutes. Within that time, every single plant in his three-hundred-acre field had been struck. Jim stood at the edge of the field and cried. He had lost $650,000.

"I had never seen anything like it," he says, grimacing at the memory. "The storm went up the valley, turned around, and came down again on the other side. The farmer whose field was right next door to mine didn't get hit at all."

It had been twenty years since his father's death. Jim was close to paying off all of his land and feeling confident about his farming future. Then the hailstorm hit. He would have to sell his land to cover the costs of the year's crops as well as his other farming debts. With the exception of his first twenty-five acres he bought in Salem in 1960, he sold it all.

"After that, I quit. I went into brokering the chile and left Faron to run the farms," Jim says. "Now, Faron and I rent farms. Most of the old farmers' kids have gone off someplace else, and they don't farm no more, so they lease the land to us. Land costs about $5,000 an acre to buy, but we can lease it at $150 to $175 an acre. This way, if you get a piece of land that's not working, you can always turn it back to the landlord."

Jim suddenly stops talking and looks sideways at Faron. "I guess you really can't blame those kids for leaving," he continues. "You have to be a really good manager to make it, and it takes a lot of money. Borrowing is tough. It's tough to be able to repay it. It's tough to buy equipment."

It is as if Jim halfway expects his son to go the way of all those other farmers' sons, as if he is trying to tell Faron to escape the farming life now while he is young and free of too many debts. He knows that Faron could take his wife and three daughters to Las Cruces or some other city where the bugs and the weeds and the weather don't matter, where the paycheck is certain. He wants his son to know that should he ever decide to leave, his father will understand.

Faron doesn't budge. He just stands there, silent, with his work boots firmly planted on the ground. He has heard this speech before. Faron smiles his father's smile, then reaches out and gives Jim a slap on the back. Jim returns the slap and pulls Faron close to his side. "Faron's young, and I still call a lot of the shots," he says, "but he's getting better every year."

Lytle Farms relies on migrant farm workers such as Juan Contreras, bottom right, throughout the season.

The year Faron Lytle officially became foreman of his father's and grandmother's farms—1992—was a record year for chile in New Mexico. The state's chile growers harvested 34,500 acres of chile, a 16 percent increase from the previous year's record crop. Faron, Jim, and June had three hundred acres of land for growing chile between them that year. Although Faron oversaw the production of all the chile on all the farms, the final product would be sold under three distinct name brands: his grandmother's "Solar Farms," his father's "Big Jim Produce," and his own "Chile Connection."

Most of the Lytles' green chile is sold fresh and roasted. Their red chile is ground into powder or tied into ristras for use in cooking or as decor. With Jim acting as broker, contracts are signed with restaurants and processors long before the season's seeds go into the ground. The rest is sold to individuals who travel to Hatch each year to buy chile from the Lytles by the gunnysack. Finally, more than twenty-seven varieties of the Lytles' certified seed is shipped to Morocco, Germany, and other parts of the world.

At harvesttime, red chile covers the roof and sides of the Hatch Chile Express, which Jim Lytle's wife, Jo, opened in 1987. Besides fresh red and green chile from Lytle Farms, the shop features chile cookbooks, specialty chile products, and chile kitsch.

By 1992, however, with more chile being grown by more farmers throughout the state, competition was growing stiff. As chile gained national cachet during the seventies and eighties, more farmers recognized its potential as a cash crop. As demand for the product increased, so did supply. Thus, the 1992 market was saturated with red chile, forcing farmers to cut prices and leaving the Lytles with a warehouse full of red chile. Fortunately, their losses on the red chile crop were offset by the fact that, with so many people growing red chile now, they could command premium prices for their green. Faron planted three hundred acres of chile in 1993 with the intention of harvesting most of it green. Productionwise, it was a very good year, but as other farmers saw that the money was in green chile now, the 1993 market became a green chile flood.

Prices plummeted, in some places going as low as $4 a sack, compared to the standard $12 to $15 price. Much of Faron's green chile sat in the fields and turned red, and he added that to his red chile surplus from the previous year. Faron had little choice but to cut down the size of the 1994 chile crop. He decreased his chile acreage by a third, to one hundred acres in all.

"It's not always fun farming," Faron says. "Some years, about the middle of August, if my yields are down or the market is flat, I'm wishing I had a job. This year, I told my banker, as soon as I get my money, I'm going to Vegas, and in thirty minutes, I'll know if I made any money this year."

Faron picks a Big Jim chile and cracks it open. "I may have better odds in Vegas, but I guess I like the adventure here," he says. "I like the tradition."

It is Labor Day weekend 1995. Another year, another chile season is nearly through. Jim and Faron are sitting outside the Hatch Chile Express on Sunday morning. The white, rectangular building is trimmed in turquoise and pink, and this time of year, its pitched roof is covered with green chile slowly turning red. Multicolored ristras drape the storefront like half a rainbow. Inside and out, the shop smells hot.

To the left of the building, gunnysacks packed with green chile are stacked beneath a shady wooden lean-to. Each sack is cinched with a green, yellow, orange, or red "Big Jim Produce" tag that denotes the mild, medium, hot, and extrahot fruit. Customers pick their chile here by the sack or inside the store by the pound. The chile is roasted in propane roasters that Faron built and then handed to the customer hot in a sturdy garbage bag. Jim and Faron sit surrounded by the sweet smell that wafts off the roasters.

"When I put one of those tags on that chile, I'm putting my daddy's name on it," Jim says. "You're not going to get no junk, no trash in that bag."

Judging by the number of cars in the parking lot and the activity inside the shop, it looks like an abundant chile year for the Lytles. Still, the picture is deceiving. The widespread appearance of the Curly Top virus ravaged chile crops throughout southern New Mexico. Of 225 acres planted between June, Jim, and Faron this year, only eighty acres survived.

"It's this doggone funny weather," Jim says. "If you have a mild winter, the cold don't kill the bugs off. There was a lot of wind but no cold. We've had three or four mild winters in a row now, and the bugs have just built up."

In years like this, a farmer needs something to fall back on. June has her certified seed. Faron has his onions, which he sells under the label "Rose Queen." And Jim has the Hatch Chile Express, the one-stop chile shop that his wife, Jo, opened in Hatch in 1987. The store not only provides an economic fallback for the Lytles but it also has spawned various cottage industries in the area. Throughout the year, the Lytles buy chile vinegars, honey, and other local food products from producers in the area and hire local residents to make ristras. As a hobby, Jim designs the shop's decorative ristras, wreaths, and other chile decor. His original "Fiesta Ristra," a two-foot-long arrangement of flaming orange, red, and yellow pods, is one of the most popular items in the shop.

"I feel that, sooner or later, there's only gonna be three or four big farmers in this whole valley, and it will be those farmers who can diversify their crops, their products, and their marketing strategies who will survive," Jim says. "I think it's going to be people like me and Faron who are going to keep it alive."

If a visitor is not mistaken, Jim has slipped his rose-colored glasses back on. He and Faron look at each other, shake their heads, and laugh. Despite their setbacks, recent and past, they look forward to the next spring, to the day when the blades of their tractor will cut through the cold, crusty earth again. Then, as they have every year of their farming lives, they will scatter some of Joseph Franzoy's and Big Jim Lytle's chile seed upon the ground and hope that it will grow. And for that moment, they will remember that chile is not just about surviving the future, it is about celebrating the past.

"You ever had green chile and Spam?" Faron asks out of the blue. "It's a part of that old Lytle family tradition."

LYTLE'S GREEN CHILE AND SPAM®

- 1 12-oz. can Spam®
- 1 medium onion, chopped
- 2 cloves fresh garlic, chopped
- 2 cups Lytle's green chile, chopped
- 1 pint canned tomatoes
- 1/2 tsp. cumin
- 1/2 tsp. oregano
- 1 tsp. pepper
- 1 tsp. salt

Brown Spam, onion, and garlic. Add chopped chile and tomatoes to mixture. Add spices. Let simmer for 20 minutes.

Red or Green?

A southern New Mexico chile pod, left, has a sloping, almost round, shoulder separating the chile from the stem. A northern New Mexico chile, right, has a flat, or square, shoulder.

Red chile should be harvested when pods reach their maximum color. Typically, the red chile harvest begins in early October, when pods are partially dried on the plant. It continues as long as the weather is good and the plant produces high-quality pods, though pods that are harvested late in the season often have a higher incidence of sunburn and disease.

Fresh green chile is a highly perishable commodity. It has a high moisture content but loses water very quickly after picking. If harvested pods are not cooled within one to two hours, they start to get soft. If they are not refrigerated or frozen within a few days, the pods shrivel. Finally, their skins turn dark and the stems rot.

When cooked into savory sauces, either red or green chile brings a flavorful element to any meal. In New Mexico, diners often find it so hard to decide whether they want their food served with red or green chile that they ask for "Christmas," a colorful phrase that chile-lovers know means both.

Santa Fe Farmer's Market

Celia Leyba, standing in a field of chile at her Embudo farm, began selling homegrown produce at the Taos Farmer's Market in the 1950s.

The majority of chile grown in northern New Mexico each year is sold at fresh market venues such as the Santa Fe Farmer's Market. Established in 1976 as a sales outlet for area farmers, the market today helps support more than one hundred farmers from fifteen northern New Mexico counties and is one of Santa Fe's most successful summertime events. All summer long, as various crops come in and out of season, shoppers flock to the outdoor marketplace for the opportunity to buy fresh from the farmer.

In the late 1950s, Fidelino and Celia Leyba of Embudo began selling their homegrown produce at the farmer's market in Taos. Today their daughter Eremita Campos and her daughter Margaret attend the twice-weekly Santa Fe gathering. There, beneath a blue-and-white-striped canopy with an *Algo Nativo* ("Something Native") sign in front, the women sell extraordinary varieties of eggplants, tomatoes, herbs, and other produce. They also sell chile: native, poblanos, habaneros, bell peppers, and piquin. "I grow a lot of different things," Eremita says, "but chile has a special place in my heart."

After her retirement in 1987, Eremita returned to the Embudo farm where she was raised and, with the help of her mother and daughter, started a new career in farming. Situated on the banks of the Rio Grande, Eremita's four-acre mountain farm is like a modern-day Eden brimming with vegetables and fruit. "Farming is an art," she says. "It takes a lot of work and a lot of patience, but you get a lot of beauty in return."

On farmer's market morning, Eremita maneuvers her blue van into her designated space in the Sanbusco Market Center in downtown Santa Fe. Rows of makeshift sales booths stand side by side, innovative displays assembled from wire baskets and wooden crates, sawhorses, and tailgates. These are filled with beets, eggs, garlic, plums, local honey, spinach bunches, statice flowers, chile ristras, and more.

Hundreds of customers will make their way through the market between 7 A.M. and 11 A.M. As they depart with armfuls of fresh produce and other homemade products, Eremita and Margaret return to Embudo—completely sold out.

"I come to the market because it supplements my income," Eremita says, "but more important, I feel like I'm providing a service that people really appreciate."

Eremita Campos brings her Embudo chiles and garden vegetables to the Santa Fe Farmer's Market twice a week. Bottom right, Eremita with her daughter, Margaret.

PROCESS
Hot

[Chapter Five]

The Future of Chile

Part One: A Flores Farewell

Dora's, a one-room café on the eastern end of Hall Street in Hatch, is jammed by 7:30 A.M. on a November morning. The peacock-blue dining room is a popular morning stop for the valley's early-bird farmers who fuel up for the day on coffee and eggs while discussing tractors, seed, and the weather. This cool fall morning, close to forty farmers have one topic of conversation on their minds. On the evening before, November 17, 1993, the U.S. Congress ratified the North American Free Trade Agreement (NAFTA), the most comprehensive free-trade pact ever negotiated. The agreement, which was established between the leaders of the United States, Canada, and Mexico, created the largest free-trade zone in the world. In theory, some 360 million consumers from the three countries would benefit from the availability of higher quality goods at lower prices that the expanded competition would create. Also, with the gradual elimination of trade barriers between the three countries, American, Canadian, and Mexican manufacturers and other suppliers of goods and services would benefit from greater export opportunities as well. Here in the agricultural hotbed of Hatch, however, not everybody sees it that way.

"These goddamn politicians are just like these goddamn doctors," says one pinch-faced farmer in a white "Hatch, Chile Capital of the World" baseball cap. "They think they know better than us what's good for us. They want to tell us how to live."

To these farmers, NAFTA is just another threat to their fragile way of life. While the agreement proposes to benefit the U.S. agricultural community by providing farmers with continued opportunities for export growth, it also expands opportunities for Mexico to export more fruits and vegetables—including chile—to the United States. Already, these farmers say, Mexican fruit and vegetable exports are much cheaper than American-grown produce. The agreement would only encourage Mexican

Eliseo Flores of Hatch, opposite, cofounded the Hatch Chile Festival in 1971 and sells his fresh chile there every year. Standing in a flourishing chile field, above, Flores shows off a few of his prizewinning Big Jim chiles.

growers to further undercut American prices in the American and Canadian marketplaces.

The second issue on these farmers' minds is labor or, more specifically, Mexican labor. Although the agreement does not provide for open borders or the complete freedom of movement among laborers from the participating countries, its potential impact on illegal Mexican migration to the United States is unclear. The U.S. Congress addressed the issue of illegal immigration in the Immigration Reform and Control Act of 1986. The law was expected to curtail illegal immigration to the United States by, among other things, providing a one-time amnesty program for illegal aliens. Aliens who had come to the United States before January 1, 1982, lived in the country continuously since then, spoke English, and met certain other requirements were eligible for temporary residence and a chance to eventually become American citizens. At the same time, the law made employers who hired illegal aliens liable for stiff penalties, a strategy meant to discourage new illegal immigrants from following behind. That strategy failed. Today, Mexico remains the leading source of illegal immigration into the United States.

The employment of Mexican migrant laborers has been key to the success of the agriculture industry in New Mexico, especially in the southern part of the state. But as immigration has continued to rise, the U.S. Labor Department has attempted to regulate the situation by tightening the reins on the farmers who rely on the migrant work force. Among other things, the regulations ensured that laborers worked under humane conditions and were paid the minimum wage, but farmers claim it has opened the door to a host of frivolous lawsuits and other false claims against agricultural employers. They say the government has created an atmosphere of fear for workers and employers alike, making the work force more unreliable than ever before.

By 8 A.M., the farmers have dispersed, scattering to fields throughout the Hatch Valley. About a mile up the road from Dora's, Eliseo Flores is standing outside of Flores Farms, his daughter's small chile shop on Franklin Street. He, too, is mourning the previous night's congressional vote.

"The farmer is absolutely in trouble," Eliseo says. "It's pretty hard to take all this. You look at what you've worked hard for all of your life and you see it going down the tubes. It's like tossing your farm, your family, everything you have into the Rio Grande and just standing there, watching everything and everybody drown."

Eliseo wishes to make it clear in the beginning that he has nothing against immigrants. He has just finished helping his daughter, Margaret, open the chile shop where she sells red chile ristras; red and green chile by the sack; and chile magnets, earrings, cookbooks, and other chile-related knickknacks. Now he is driving down the street to his house and farm beside the Hatch railroad tracks. "I'm not prejudiced against Mexicans," he says. "I'm prejudiced against people who don't want to work."

Although Eliseo was born in New Mexico, he admits he might be considered an immigrant himself. According to oral family history, his grandfather emigrated from Spain to Mexico, where he kidnapped an Indian maiden and made her his wife. Their son, Eliseo's father, was a soldier in the Mexican military who apparently came to the United States after being chased by Pancho Villa from Mexico to Kansas City. He eventually settled in Hurley, New Mexico, where he married and began raising seven sons. Eliseo was five when his father died in 1929.

Three years later, Eliseo's mother moved her family some one hundred miles west to the burgeoning farming community of Hatch. There she sent her boys to school and secured jobs for them on area farms. "Everybody here in the valley was so nice to us," Eliseo recalls. "We were hustlers, we worked for anything. Most people paid us with produce from their farms. It was then we started working with chile."

It was a hard life, but Eliseo loved every bit of it. He never dreamed of leaving small-town Hatch for life in a big city. His dreams were much simpler than that. Someday he wanted to buy a little piece of land on which to grow some chile or cotton or some other thing, and he wanted to marry a woman named Margaret just because he loved the sound of the name.

In 1942, Eliseo was drafted and later sent to the South Pacific, where he spent the duration of the war. When he returned to Hatch, he took a job selling furniture, appliances, lumber, and hardware. He rented a few acres of land where he tended chile and cotton before and after work. Finally, he married a woman named Margaret.

Eliseo worked at his sales job for the next thirty years, all the while maintaining some farmland on the side. He started building a home, and in 1952, bought his own farm, twenty-eight acres in Hatch. Combined with the land he rented, Eliseo worked 130 acres in all. By 1964, his total acreage had climbed to four hundred acres, providing him with enough money to buy out three other farmers and create a second forty-six-acre farm of his own. With the income generated from his acres of chile, cot-

Eliseo Flores and Babe set out for their daily trip to the Flores Farms chile fields.

Chile piquin, grown from seed that has been in the Flores family since 1932. At the 1995 Hatch Chile Festival, Eliseo Flores received ribbons for his Sandia, jalapeño, Barker, yellow hot, and Big Jim chiles.

ton, alfalfa, and hay plus his sales job, he made monthly payments on his land and acquired farming equipment. Chile had by then replaced cotton as the valley's most lucrative crop, and in 1971, he joined a group of other farmers in founding the first annual Hatch Chile Festival. To Eliseo, the festival was much more than a venue in which to promote the local chile crop; it was a statement of pride in the agricultural way of life.

"It was beautiful to be a farmer then," he recalls. "There was no competition because there was plenty of land and plenty of chile to go around. You could buy a farm and that farm would make you enough of a living so that you could actually pay it off someday. Those really were the good old days."

Like all good things, the good old days would come to an end. Eliseo says the chile industry begin to change about 1979, four years after he retired from his sales job to farm full-time. The state's chile crop was booming as chile acreage increased to fifteen thousand acres that year. By then, he had some 500 acres of land, most of it in chile, spread among eleven separate farms throughout the valley, and his own two farms were nearly paid off. He had twenty-two steady employees plus 120 more seasonal workers at harvesttime. He had up to eleven different varieties of chile growing at once to fill his contracts with canneries, dehydrators, and other processors. The rest he sold by the sack off the back of a flatbed at Hatch Chile Festival time. Eliseo ran his whole operation at the cost of about $90,000 a year. Chile prices were good, but his operating costs were starting to climb.

Eliseo is walking through his farmyard now, where eight massive tractors and all their accoutrements are scattered about a small community of metal buildings tucked between chamisa, sagebrush, and dust. "Anymore, your equipment costs more than your land," he says, addressing his tractor. At six foot tall, Eliseo's lanky frame barely hits the top of the giant tractor tire that in 1979 cost him $250. Today, he pays $800 to have it replaced. Back then, he bought the entire tractor for $32,000. Now the same tractor sells for $80,000 without the extra parts. Meanwhile, the gas to run the tractor has gone from thirty-eight cents to $1.80 a gallon and a tankful of fertilizer has risen from $180 to $700.

The superior prices that farmers were getting for their chile in the early 1980s supported the price increases for a time, but as the market became more competitive, and chile prices declined, farmers began to feel the strain. For Eliseo, the stakes got even higher when the U.S. Congress passed the Immigration Reform and Control Act in 1986.

The harvest that year was especially memorable for Eliseo. One July morning, he was picking green chile when a hailstorm suddenly swept through the field. The hail came from the north and ran southward along the length of the chile rows. Then, just as quickly as it had arrived, the storm lifted and disappeared. Eliseo stood stunned in the middle of the field with one side of his face completely numb and an earful of ice. As he started to pick the hail out of his ear, he heard a sound like a waterfall streaming through his head followed by the high-pitched whistle of a faraway train. The rushing water and the whistle rang in his ear for an entire week before a doctor determined that his eardrum had burst. In all of his years of farming, he had never had a major injury. He took his first very seriously.

"I wondered if this was some kind of sign, like it was time to stop and think about what was happening to my work and my life," Eliseo recalls. "But ever since I was a little boy, I learned that farmers always look forward to the next year being better than the last. My kind of life had always been to get ahead and make something of myself. Things were getting worse, but I still looked forward."

The 1995 Hatch Chile Festival is kicking off another Labor Day weekend, and Eliseo Flores has taken shelter from the 105-degree heat in an air-conditioned RV parked behind the flatbed full of chile he has for sale. Tacked to one of the flatbed poles, between the fresh green chile and the red chile ristras and powder, is a string of ribbons that Eliseo has been awarded for his chile this year. Two of his granddaughters proudly point out their grandfather's win: four blue first-prize ribbons for his Sandia, jalapeño, Barker, and "Yellow Hot" pods plus a yellow third-place ribbon for his Big Jim. When Eliseo emerges from the RV, however, neither his face nor his mood indicates that he is especially happy—about the festival, about the awards, about anything.

"That don't mean I'm a good farmer, that means I'm a hard worker," he says with some bitterness, indicating the awards. "There are a lot of farmers here who are much better than me, who have more money than me, but I work hard every day. I'm a jackass of all trades. I have to be because nobody else around here seems to want to work."

Eliseo shakes his visitors' hands and then launches into a ten-minute tirade about his recent problems with labor. He shows not the slightest bit of compunction as he expresses his disdain for the immigrant work force, which he claims has become utterly unreliable since the establishment of the amnesty program in 1986. He accuses laborers of stealing farm equipment, of lying to the labor board, and of using the government to drain area farmers of money without doing the work. In turn, he accuses the government of turning its back on hardworking, tax-paying, law-abiding Americans such as himself.

"These people need to work, but they're abusing the farmer and the government backs them up," he says. "We're no slave drivers. We provide the ice water and the portapotties and pay the minimum wage. We're just trying to give these people a job. Me, I'm just an old-fashioned guy. I fought for my country. I've always paid my insurance for my vehicle. Everything I have, I've worked for. I'm not asking for any handouts from anybody. All I'm asking for is help, and we don't get it."

Maintaining good hired help has become so difficult for Eliseo that his once-sizable work force has dwindled down to one. At the same time, he has let go of several of his rented farms, leaving him with land spread between two small rented farms and the two farms of his own. Of that acreage, he has drastically reduced his chile crop to this year's total of fifteen acres; the rest is planted with the much less labor-intensive crops of alfalfa and hay. Eliseo's only son, Richard, worked for his father for twelve years, hoping that someday he could take over the family farms. At his father's advice, however, Richard recently took a job with the state. Eliseo has canceled his contracts with the canneries, the dehydrators, the processors. The high physical, mental, and financial costs of growing chile have taken their toll.

"My operating costs had gotten up to about $200,000 a year, and I'm making five cents an hour. I've been abused so much, I just can't cope with it anymore," he says. "I used to be a Democrat, but I'm nothing no more. My vote don't mean nothing. Nobody hears me. In the last few years, I've even plowed up some of my chile. I'd rather plow it up and get some satisfaction that I'm the one who's calling the shots."

With that, Eliseo retreats into the cool RV, leaving his granddaughters to handle the chile sales. The encounter is brief, powerful, and intense. Within an instant, though, Eliseo is back, apologizing for his mood, regretting his words, grasping for understanding.

"You know, the only time I've ever missed farming was during the months that I was in the service," he says, "but I feel like I'm under more pressure now than when I was at war."

There is a long, sad silence, which Eliseo finally breaks.

"Meet me at Margaret's shop in the morning," he says. "Eight o'clock."

Eliseo wants to show his visitors his chile crop. He wants to show them how his story ends.

By morning, Eliseo's face has softened and his words are much lighter than the day before. He is parked outside of Flores Farms, seated on the tailgate of his white Ford with Babe, his beautiful black Labrador. When his guests arrive, Eliseo nudges Babe into the back of the truck and says, "Follow me."

It takes less than a minute to reach the outskirts of town, where a series of farm fields is connected by a meandering network of dirt roads. Eliseo takes a long, bumpy path that curves alongside a bend in the Rio Grande. Squeezed between the river and the road, wide blankets of alfalfa and cotton alternate in brilliant shades of green and white, creating a lush riverside quilt. Finally, Eliseo arrives at the chile field. He gets out of the truck and, with a dramatic sweep of his arm, declares, "From San Diego Mountain on the south to Caballo Lake on the north, this is the Chile Capital of the World." And here in its center lie Eliseo's fifteen acres of chile in neat 1,200-foot-long rows.

The field is flourishing. Prize-winning Big Jim plants stand thirty-six inches tall with pods so large it takes only three or four to make a pound. Further downfield, the rows of Big Jims give way to blue-ribbon Sandia, jalapeño, Barker, and Yellow Hot. Eliseo walks through the field to a small corner section filled with chile piquin, a tiny bright red pepper that is the contemporary counterpart to the original chiltepin chile that first grew wild in South America nine thousand years ago. The seed that was used to sow this crop has been in the Flores family since they arrived in the Hatch Valley in 1932. Eliseo grows it every year. He says it reminds him of where he came from.

"I'm real pleased with all of my chile. I could never put my chile down," he says. "I know that it's my chile that has made me the money I've made, and I know it's the government that has taken it away."

Eliseo planted this field of chile on his birthday, March 4, hoping it might bring him a little extra luck. Despite two weeks of spring frost and fifty-mile-per-hour winds, the chile survived. It not only survived the weather but also the vicious Curly Top virus that destroyed crops throughout the southern part of the state. Still, there is not enough hope left in him to make him think that his crop's survival might be a good sign for the future. Eliseo announces that he has made a decision: He is selling his forty-six-acre farm.

"I'll be honest with you. The reason why I'm selling most of my land is because there's nothing for me to look forward to anymore," he says. "I once looked forward to leaving it for my son and daughter and my grandkids, but I don't want my kids or my grandkids to be farmers in this day and age. I won't have nothing to do with these government regulations anymore. I've had enough."

Eliseo is not the first farmer to give up on government, and he surely won't be the last. But after a lifetime of working the land, he is too much of a farmer to completely give up on farming. He says he will keep his small farm and some of his equipment, and every year he will grow a few rows of chile, just enough to keep his daughter going in her chile shop. He'll plant it, water it, weed it, and pick it himself. After more than sixty years of farming, Eliseo is still not afraid of hard work. By 7:30 this morning, he had already picked a truckload of chile. By the end of the day, he will have filled two hundred sacks.

The prospect of retirement clearly makes Eliseo uncomfortable, and one wonders if he will really go through with his plan. If he does, one wonders if his anger can ever be transformed into the acceptance that even the agricultural way of life will change. But Eliseo is a proud man, and after a moment, he starts to consider the positive things that retirement will bring. He'll have time to go hunting and fishing. Maybe he'll even go on vacation once in awhile.

"Do you believe my last vacation was in 1976?" he asks. "I think I deserve the rest."

There in the middle of the field, with his bushy chile plants climbing all around him, Eliseo looks almost content, for despite all his struggles, he has at last become everything he ever wanted to be: a small-town man with a simple, small-town life. A man with a little piece of land and a wife named Margaret.

Part Two: The Mother of Invention

The mechanical chile harvester that Gary Riggs purchased strips the chile pods from the plants.

There is a rattle, then a roar, and finally a mesmerizing hum that blocks out all other sound. Gary Riggs of Garfield is seated at the controls of his green-and-red "McClendon Pepper" chile picker, smiling. Slowly, deliberately, it crawls across the forty-acre chile field, passing over two rows at a time, magically, mechanically, stripping the chile plants of their ruby fruit. It takes about five minutes to reach the other end of the field, and by the time he gets there, Gary's smile is even wider.

"This right here," he says, indicating the chile picker, "this is the future of chile."

For some farmers in New Mexico, the future of chile is looking a little grim. In the southern part of the state, problems with labor and increased competition have left some struggling to define their place in the industry. Growers in the north battle the perennial risks of a short growing season, which often results in a less-than-profitable crop. Everywhere, the increased risks and costs of farming stir questions of whether or not the next generation would, or even should, take their parents' place on the family farm. For everyone, everywhere, there is an agricultural tradition at stake. Some people will quit the business altogether, relegating that tradition to sentimental memories of the good old days. Others will remain, solely for the sake of tradition, whether they profit or not. Still others will look for alternatives, hoping to find a middle ground where tradition and success can coexist.

Inherent to that success is the question of how to pick chile in the most efficient way, a question that Fabian Garcia began pondering when he started his chile research at New Mexico State University in the early 1900s.

Gary Riggs found his answer in Tulia, Texas, a few years back when he bought the pepper picker. At a cost of $115,000, it wasn't cheap, but labor problems had left him with no alternative. In 1992, Gary lost an $80,000 lawsuit after a group of farmworkers claimed that he wasn't paying them the minimum wage they were guaranteed by law. Gary, who had hired the employees through a labor contractor, argued that it was the contractor who was not paying the employees in full. The court disagreed.

"It was a frivolous lawsuit, but I lost, and I even had the Labor Department on my side," Gary says. "Things had just gotten way out of hand. I was totally depressed with the whole situation. I had to find another way."

The Riggs family has had mechanical harvesting on its mind for a long time. Gary's grandfather, Fred Riggs, Sr., raised chile in the Hatch Valley communities of Garfield and Arrey, and his father, Fred Riggs, Jr., followed the same path. "My father's forgotten more about chile than I know," Gary jokes. He was ten when his father, grandfather, and uncle Ernest Riggs built a dehydrator to dry their red chile. Next, they set out to build a chile-picking machine. It was a tricky endeavor: constructing a machine that would remove the fruit from a plant without damaging either the pods or the plant. Even trickier was getting the machine to distinguish between a mature and an immature pod. The picker never worked to their satisfaction.

All these years later, Gary is one of the first farmers in the Hatch Valley to use a chile picker to harvest his crops. A few others are experimenting with their own homemade machines, but his picker is one of the few that has seen some success in the mainstream marketplace.

Gary estimates that the machine can pick up to four acres of chile a day, depending on the condition of the field. In terms of labor, that translates to the work of about forty people. Unlike human harvesters, however, this machine cannot determine the difference between a mature and an immature chile pod. For that reason, he still employs some laborers to help the process along.

This morning, pod maturity is not a question as the machine moves through a field of ripe Joe E. Parker pods destined to become red chile powder. Looking down from the glass-enclosed cab, one sees a set of two helices, each one poised directly above a row of chile plants. As the machine moves forward, the rotating helices grasp the pods by the stem and pull them off of the plant. "It's set close enough to where each stalk has to go through it," Gary says as he guides the picker. "You can't miss."

Meanwhile, another man drives a tractor directly alongside Gary. He tows the eight large wooden crates where the chile is deposited after it travels up a conveyor belt and drops off the picker. Six other workers walk behind the picker, rescuing any pods that might have fallen on the ground. Another stands knee-deep in chile atop the wooden crates, making sure the pods are free of extra branches or leaves before they can be taken to the processor.

After a few passes through the field, chile pods have splattered on the glass of the cab like bugs stuck to the windshield of a car. The view from the back window shows the machine has made a clean sweep: two rows of empty chile plants lie in its wake while the two rows beside them still overflow

Gary Riggs, top right and father Fred Riggs, Jr., have looked forward to the age of mechanical chile harvesting for many years. Bottom: Row after row of chile plants lose their fruit as Riggs's pepper-picker makes its way through the field.

with red fruit. The first eight crates are filled in about an hour, and eight more arrive. As another tractor drives the new crates into place, Gary shuts down the machine for a moment's rest.

"I've been farming about fifteen, twenty years, and this is one of the best things to ever come along," he says. "We can use this machine to harvest jalapeños and other kinds of chile, too. We've got 140 acres of chile this year, and we'll have about 250 acres next year. I'd like to get up to about 450 acres a year, even more. We just might be able to do it now."

Gary waits a few minutes and then restarts the red-and-green beast. The future is right in front of him. He can see it through the chile-stained glass.

Back in the early 1980s, Las Cruces chile veteran Robert Cosimati was finding that picking clean, mature chile by hand was no longer cost-efficient. He was finding too many leaves, plant limbs, and dirt clods mixed among the harvested pods, some not yet mature enough to have been picked to begin with. Immature pods and excess plant residue are offenses for which a farmer is docked at the processing plant. In addition to his laborers' daily wages, he was losing money on their poor pickings. "I said to my wife, 'What we need to do if we're to stay in the chile business is to figure out some way to pick mechanical,'" Robert recalls. "I said, 'The industry has either got to go push-button or go out of business.'"

Instead of relying on the industry, Robert and his wife, Barbara, decided to do the button-pushing themselves. Fifteen years later, the "Crown Chile Harvesting System," a bright white automated chile picker, is parked in the front drive of their Las Cruces home. The gargantuan creation is Robert's own invention, the culmination of his years of labor frustrations, farming experience, and mechanical know-how. It is his best substitute for human hands and his last hope for the future of chile.

Robert and Barbara are sitting inside of their large, cool living room with a thick scrapbook open on the coffee table before them. The book documents the evolution of Robert's chile picker in photographs, with their chronology and other important details noted in Barbara's handwriting underneath.

"I wasn't born into a farming tradition," Robert begins.

"He's a city boy," Barbara jumps in. "It was me who made him into a farmer."

Robert was born in Albuquerque and lived there until 1947, when his parents, Ozzie and Della, moved their family to Las Cruces. His father, who was one of the original founders of Southwest Distributing, then a wholesale liquor distributor, had come to run the company's Las Cruces branch. Robert was suddenly surrounded with agriculture, but he had no particular interest in the land or its life-style. He did, however, develop an interest in a farmer's daughter. Her name was Barbara Taylor, and he married her in 1956.

For a time, Robert followed in his own father's footsteps; the couple moved to Albuquerque where Robert drove one of the distributorship's liquor trucks. But Barbara wasn't much for the city, and the couple eventually returned to Las Cruces, where Robert took a job in a men's clothing store. In 1958, she convinced her husband to take a shot at working on her father's farm. Robert endured two years of backbreaking farmwork then went back to selling men's clothing. This time, however, he lasted only eight months.

"I thought, 'Yeah, fitting suits is a lot easier than farming, but it's not any fun.' I came back to the farm because I realized I enjoyed it. Hard work isn't so bad if you enjoy what you're doing."

Robert worked with his father-in-law, Anzlie Taylor, growing cotton, alfalfa, and chile for fourteen years before buying him out of his farming equipment and acquiring his land leases in 1972. Before long, he had nineteen landlords and was farming up to 1,400 acres throughout the Mesilla Valley. The harder he worked, the deeper he fell in love with farming.

"It really is kind of romantic," he says. "You go out and plant a seed and it grows. I can't tell you what kind of satisfaction there is in opening up your door and seeing a field full of chile in bloom."

By the summer of 1981, Robert had two hundred acres of chile blooming across the valley. In terms of the crop itself, it was a great time to be growing chile. "It was a good cash crop, we were getting very good production, and we had very few viruses," he recalls, but the fact that it was such a labor-intensive crop made it difficult to manage, especially at harvesttime. First there was the bookkeeping and other paperwork. Every morning before they could begin work, the laborers had to sign in with Barbara and have their documentation papers checked to ensure that they were legal immigrants. At the end of the workday, the number of buckets each worker picked was tallied and recorded and the worker paid in cash. If for some reason a worker's final daily wage did not work out to the minimum wage, Robert was required by law to make up the difference.

Even worse than the bookkeeping were the clods of dirt that Robert was finding in his chile at the end of the day. He was working eighteen-hour

Robert Cosimati, of Las Cruces, shows off the Crown Chile Harvesting System, his homemade mechanical chile picker.

Cosimati says he and his wife, Barbara, designed their harvester to work from the farmer's point of view.

days and spending upwards of $2,000 a day for every fifty laborers he hired to harvest his crop. He was beginning to lose not only money but with it his feeling that farming was fun.

In 1988, a University of Michigan researcher reported the number of chile harvesters built in the United States at 126. Those on the market were distinguished by three different picking apparatuses. The first type picks with angled or vertical open double helices, a system similar to Gary Riggs's machine. The second is equipped with rubber fingers and brushes that remove the fruit from the plant. The third employs a propeller system in which revolving metal bars slice the pods off the plant at the stem. Robert's picking system was meant to be different from them all.

"I had seen some machines that a few people were trying to build, and I looked at some from other companies," he recalls. "But I thought, 'If I'm going to spend $100,000 on a machine and not be sure it's going to do exactly what I want it to do, then why not spend that much money on my own?' I wanted a chile picker that was made from the chile farmer's point of view."

Robert was no mechanical engineer, but he had a penchant for "piddling," which he often put to work building remote-control model airplanes. So, as soon as the disappointing 1981 harvest was through, he began piddling his way toward his first chile picker. He started with a John Deere haycutter, which he stripped down to its chassis. Then he added a spiral picking unit and sixteen dashboard variables that adjusted for different crop conditions and speeds. He tested the machine in the field for the next four years, making modifications every step of the way.

"It was a mess," Robert recalls.

"It was a very crude method," Barbara adds, "but it was still better than people."

Robert got so disgusted with the picker that he parked it for two years. When he undertook the project again, he adapted a picking propulsion unit from a John Deere cotton stripper to fit the machine. Next, he added a revolving barrel cleaner, a tool commonly used by processors to remove leaves and other residue from the pods. He tested and modified the picker for another four years. Each time he came up with a new process or mechanism that worked, he applied for a patent, eventually receiving nearly thirty patents in all. The machine was getting better, but it was far from perfect. In 1991, Robert collaborated with a farmer in Hatch, who designed new picker heads for the machine. The heads were intended to pick a cleaner product.

"It was a process of elimination," Barbara says. "It was all put together by hand with what was available."

"I was trying to design it to where it would be repairable in the field," Robert says, "so that as long as you've got a machine shop and a John Deere dealership around, you could fix it yourself and didn't have to depend on some distributorship."

Finally, in 1994, Robert deemed the chile picker complete. Standing twenty-seven feet long and nearly fifteen feet tall, the convoluted mass of metal could do everything he intended it to. It picked two rows of chile at once and could be adapted to fit thirty-six- to forty-inch rows. His original machine picked only red chile, but this picker harvested mature green chile and the smaller jalapeño peppers as well. The machine could be set to pick the pods at the bottom of the plant, where the pods reach maturity first, while leaving the immature pods at the top to grow. It could pick the same amount of chile as fifty laborers could harvest in a day, and by the time the pods had gone through the machine, they were clean enough to take directly to the processor.

"I designed it so that everything that comes out of this machine is processible," he says. "This chile is clean."

So clean that Robert made the machine to be lubricated with cooking oil instead of motor oil.

"I'm trying to give you a clean product. You're going to eat this, after all," he continues. "Cooking oil won't harm the chile. You can wash it off. If you have to dump some in the field, it won't hurt anything."

So clean, in fact, that Robert painted the picker white.

"I picked the cleanest color I could find. All the other machines out there are red or green. When you see this machine, you'll know it's mine."

Barbara slaps the scrapbook shut, and she and Robert lead their visitors outside to examine his chile harvesting system. The picker towers above the house, the storage sheds, the trees. A clear-glass cab equipped with a steering wheel and a dashboard full of controls sits directly behind four spiral pickers. A complex series of conveyer belts and other large metal devices bring up the rear.

"That's a darn good-looking machine," Barbara says. "I'm proud of that machine."

In any other setting, this contraption would appear completely out of place, but here in the heart of the Mesilla Valley, its seemingly random mechanical arrangement makes perfect sense as it glides through a field with accuracy and grace. Traveling up to five miles per hour, the front end of the picker moves along the chile rows, its front-end spirals detaching the pods at the stem. As the pods travel through the last spiral, they are dropped onto a conveyor belt that carries the chile to the rear of the machine. From there, the fruit travels along two more conveyor belts and into a blower that whisks excess leaves and dirt off the pods. It takes two more conveyor belts to deliver the pods to the barrel cleaner, which, spinning at variable rates, rids the chile of remaining debris. Finally, the pods are thrown out of the barrel and onto one last long conveyor belt that spans the length of the machine. There, three to five laborers sort and examine the pods before the fruit is dumped into a trailer at the picker's rear.

"You will never beat the human eye in making sure that the pods are coming out of the field right," Robert says. "But for the most part, this machine doesn't go on strike. And it doesn't sue!"

Except for the few workers he hires for quality control, he has not employed a chile harvesting crew since 1993. The displacement of the migrant work force due to increased mechanization in the industry is an issue that is now beginning to be discussed, but Robert believes that, ultimately, mechanization not only will be better for the industry but for the laborers as well.

"Where are the laborers going to go?" he asks. "It's just like it was when the cotton picker came in the 1950s. They're going to get better jobs and make more money. They will no longer be tied to migrant labor because their job won't exist anymore."

In terms of the industry, Robert has so much faith in mechanization that in 1996, for the first time since he started farming in 1958, he didn't plant any chile of his own. Instead, he will hire himself and his chile machine out to harvest other farmers' crops. Once these farmers see the results his machine achieves, he believes he can start convincing them to buy one of their own.

"A lot of them think it's great, a lot of them laugh at it, but that's what I expected," he says. "But I can prove to these farmers that if you have a hundred acres of chile, you can pay for this machine in three seasons. You can't do that with fifty laborers."

At the 1996 price of $140,000, the Cosimatis had yet to sell even one picker. "We've learned a lot, and it's cost plenty," he says, "but I don't want farmers to buy this until it's 100 percent perfected. I want to know that when a farmer buys it, it's perfect, so they can be proud of it."

Robert estimates that it will take another $500,000 to make his machine marketable. Already, he has a Las Cruces company ready to manufacture the pickers. If the propulsion system is readily available, he says the company can build one every eight weeks.

"It's an ongoing thing. We're in the Model T stages here," he says. "This will constantly be improved. As more technology comes about, we'll add to it."

"But we're already before our time," Barbara adds. "The industry is just going to have to catch up with us."

Ultimately, Robert explains, farmers will have to rely on mechanization, while consumers will have to be willing to pay higher prices to support the rising costs of raising the crop. Even chile itself must change.

"What we need from the chile breeders at New Mexico State University is a plant designed for machine harvest that grows rapidly and sets all at one time," Robert says. "A twenty-first-century chile, if you will. We need to keep all the centuries and generations that have gone into the chile crop alive."

Elizabeth Berry

Elizabeth Berry grows chiles from around the world at her three-hundred-acre Gallina Canyon Ranch in Abiquiú.

The striated stone cliffs of Abiquiú in northern New Mexico are perhaps best known as the land of Georgia O'Keeffe. But since 1980, Abiquiú has also become known to many as the farmland of Elizabeth Berry, a California transplant who has cultivated some seventy-five international varieties of chile in her high-desert locale. Working between her 300-acre Abiquiú canyon ranch and a ten-acre farm in the bottomlands below, Elizabeth has forged a special niche for herself in the burgeoning chile business.

"If you're going to grow anything in New Mexico, chile is the obvious thing," Elizabeth says. "But I don't grow New Mexico chile. I try to grow what the others don't grow."

In fact, Elizabeth did not come to New Mexico to grow chile at all. Shortly after arriving in Santa Fe in 1977, Elizabeth met Mark Miller, an acclaimed California chef who had moved to Santa Fe to start a restaurant. Miller wanted to elevate traditional New Mexican foods to southwestern nouvelle cuisine by incorporating chiles from around the world. While searching for a local chile source, he urged Elizabeth to try out her green thumb.

"Mark told me what kind of chiles he wanted, so I got a seed catalog and proceeded to look them up," Elizabeth recalls. "I didn't know what I was doing. I just did it."

Perhaps the biggest edge that Elizabeth has in the chile business is the type of seed she grows. A strict disciple of "heirloom" agriculture, she shuns hybridized chile seeds in favor of those that are as close as possible to their original form.

By the fall of 1995, Elizabeth was supplying specialty chiles to Miller's highly successful Coyote Café, as well as to some sixty other chefs in Albuquerque, Santa Fe, and Taos. Among her exotic repertoire are chiles from Mexico, Italy, Brazil, Hungary, Korea, the Caribbean, and Peru, chiles that run the gamut of heat, color, shape, and taste. One particular Peruvian *aji,* for instance, has the same tangy flavor as a lemon slice. The Mexican ancho chile, which is the dried version of the Mexican poblano pepper, is smoked by Elizabeth over basil wood to bring out undertones of licorice, raisin, and plum. Elizabeth's biting *de arbol* chiles from Mexico are used by chefs to spice up sauces, soups, and stews, and at least one bartender substitutes one of her deep purple chiles for a martini olive.

A *sampling of Elizabeth Berry's exotic chiles creates a mosaic of color, form, and taste.*

Chapter Six

Hatch, New Mexico: Chile Town, U.S.A.

The Rio Grande doesn't look so large to a traveler turning west from Interstate 25 onto State Highway 26 headed into Hatch. Crossing the water via a squat, paved, two-lane bridge, the river appears scarcely wider than a hundred feet as its borders wind beneath the bridge like a slow-going snake. Along with the northerly Caballo and Elephant Butte reservoirs, however, this stretch of the Rio Grande feeds some of the most fertile farmland in New Mexico. From the fields of the Hatch Valley—from Arrey to Derry, Garfield to Salem to Rincon—come crops of onion, cotton, alfalfa, lettuce, and pecan, crops so historically prolific that, in 1959, the Hatch Chamber of Commerce adopted "Never a Crop Failure" as the valley's agricultural slogan. Of course, many crops had failed before that, and many have failed since. Still, the chamber optimistically clings to its claim; as one representative put it, "It gives farmers something to hope for."

The Rio Grande provides a picturesque entryway to the small southern town of Hatch.

Like oranges in Florida or potatoes in Idaho, there is only room for one agricultural celebrity in these parts, and it is chile. Some of the most productive chile farms in the country fill this arable slice of southern New Mexico, giving the chamber of commerce reason to claim another venerable valley motto, the "Chile Capital of the World." Botanically speaking, there is no such thing as a "Hatch" chile. Nonetheless, Hatch is the most recognizable name in New Mexico chile, so popular that people have been known to sell chile grown elsewhere under its name, risking arrest and a stiff fine.

The town of Hatch itself, a country hamlet set in northern Doña Ana County in the heart of the valley that bears its name, gets the most mileage out of the valley's claim to fame. "Hatch, Chile Capital, Welcome," screams the bright red-and-yellow sign that greets motorists who have just crossed the bridge over the Rio Grande. Except for all the promotional signage plastered on storefronts and truck bumpers throughout town or the sight of chile roasters and chile ristras that fill the streets each fall, though, there is not much within the four-square-mile village limits to distinguish Hatch as a modern chile mecca. Drive into town in the middle of May, when perfect rows of onion bulbs are busting through the earth and the roadside smells like a simmering pot of onion soup, and you may just as well be in the capital of onions. Most of the time, when it is not busy being the Chile Capital of the World, Hatch, population 1,100, elevation 4,050 feet, is just another small town.

Opposite: A lonely length of railroad track is a reminder of the late 1800s, when Hatch was a bustling stop on the Atchison, Topeka, and Santa Fe Railroad.

Robert Duran, a former three-term mayor of Hatch, enthusiastically recites the virtues of living in Hatch as if he were a salesman describing the features on a used car.

"It's not an expensive place to live, you can find a lot of homes here, taxes are pretty low," he begins. "The people are wonderful people. Our school system is central, easy to get to. We have one of the best medical clinics in the state. We're fifteen minutes from where you can go hunting deer in the Black Range, a beautiful forest. We're near two of the most beautiful lakes in the state of New Mexico. We're only thirty-six miles from Las Cruces, the second biggest city in the state. We've got four Catholic churches, nine Protestant. Oh, and the weather is great. Never too cold or awfully hot."

According to a chamber of commerce brochure about the village, which the former mayor handily produces, the average local temperature in Hatch is 79.1 degrees while the average rainfall is 8.96 inches a year. The former mayor muses that it must have been the comfortable climate that attracted the original settlers to the area, beginning with a large group of Indians who had lived in the Jornado Range for centuries before the Spanish arrived in 1598. The route from Mexico City through El Paso into New Mexico followed by the Don Juan de Oñate expedition in that year was known as *El Camino Real*, the Royal Road. It passed a few miles east of present-day Hatch as it wound northward along the Rio Grande to Santa Fe.

Some 250 years later in 1851, a group of New Mexicans settled a small village in the area and called it Santa Barbara. Subsequent settlers named it after General Edward Hatch, a former commander of nearby Fort Thorn. Hatch remained small and isolated until the early 1890s, when the Atchison, Topeka, and Santa Fe railway built a flag station there alongside the path of the railroad's diagonal line between Rincon and Deming.

The village of Hatch was incorporated in 1927. By the time the chamber of commerce coined its "Never a Crop Failure" slogan in 1959, acres of

chile filled the surrounding valley and the town's population topped a thousand. Twelve years later, a group of local businesspeople and farmers joined forces to create the first annual Hatch Chile Festival. Held on Labor Day weekend at the peak of the annual green chile harvest, the festival was conceived as a way to draw New Mexico visitors and residents alike to Hatch to purchase their chile stock. The event, which took place at the Hatch Airport on the southwest edge of town, featured dirt bike races, horseshoe pitching, and contests for best beard and best legs. Various local farmers received ribbons for growing the most perfect green chile pods in the valley while "Little Miss Green Chile," "Little Miss Red Chile," and a "Hatch Chile Queen" were crowned. About 1,500 people attended the first festival. The "Chile Capital of the World" was born.

Duran runs his finger up and down a map in an outline of the Hatch Valley, dividing the local employment picture between the schools, the medical clinic, the dairy, the state prison in Las Cruces, and the White Sands Missile Range. He hesitates when he gets to farming.

"Chile is one of the things that has paid best in the last twenty-five years, but you have to remember that a farming community is usually a poor community," he says. "Most of the farmers live outside of the community, and the village doesn't make any revenue from their sales. All that the village gets are the gross receipts taxes that are left in the community by the farmer, and that's not very much. We have a good chamber of commerce that gives us the chance to have the Chile Festival and be the Chile Capital of the World, but this town would have died a long time ago if we weren't at the crossroads of Interstate 10 and Interstate 25. If it wasn't for that, Hatch probably wouldn't survive."

Per capita, Hatch must have one of the highest percentages of trucks and tractors in the world. From the outlying farmhouses to the residential neighborhoods that span the village interior, there is at least one truck and tractor in every driveway, usually two, quite often three. When not parked, they are ambling in all directions along the roadways, ducking into farm fields, then climbing back onto the road again. In the village proper, one sees them at every turn. Early morning to late afternoon, they turn into the multiple hardware and auto parts shops that dot Franklin and Hall streets. You can see their reflections in the large display windows of the Hatch Mercantile Company, where pantyhose and diapers share space with boots, blue jeans, and enamel pots. They pass the mercantile on their way to Rosie's or Gloria's or the local Dairy Queen, where the charbroiled burgers are served with a pile of Hatch green chile on top.

Johnnye Hammett has seen a lot of changes in the village since she came to live there in 1948.

After awhile, one learns the best way to ask directions in Hatch is to ask the color of the truck that is parked in the drive. Johnnye Hammett proves the exception.

"It's number 713, I believe. We don't use those numbers too much out here," Johnnye says over the phone when asked directions to the Hammett house. "It's the one with the fancy sports car out front."

The 1995 Chrysler Sebring is parked beside an immaculate green lawn and a white wrought-iron sign that says "The Hoot Hammetts." Johnnye appears and explains her unusual name: "My mother loved me fiercely, but she wanted a boy, so she named me after my grandfather. My real name is John."

John Franchie Harriett Culverson Hammett is the second woman ever elected to the village of Hatch's board of trustees. During her tenth and final year in office, Johnnye held the position of mayor pro tem.

"I'm the first lady to crack that one," she boasts, "but I'm not really your typical old lady. Why, I just bought myself that fancy sports car."

Johnnye's living room is populated by her collection of stuffed, ceramic, wooden, and pewter pigs. The walls are covered with photos of her two children and her late husband, Leon ("Hoot") Hammett. A picture of Hoot's namesake, the western movie star Hoot Gibson, occupies a spot on another wall. Hoot and Johnnye had been married three years when they came to Hatch in the summer of 1948. Hoot took a job with the El Paso

Electric Company, while Johnnye began a thirty-year career in the finance office of the Hatch Valley Municipal Schools. When Hoot died in 1986, Johnnye had just been elected to her first term on the village board of trustees.

"Hatch is a country town," she explains, "where seniors can drive, where you can have a good little group that plays dominoes on Thursdays or bridge on Tuesdays, where the ladies like to exchange an apple pie for a chocolate or a pecan, where everybody has their church. My favorite Hatch fact is that there are no more people in Hatch now than when I came here in 1948, and no fewer, either. Of course, now we have a lot of folks from Mexico. That may sound race prejudice, but it isn't. The complexion of Hatch has changed."

As is the case in conversations throughout town, Johnnye's dialogue almost naturally has drifted to a topic as big in Hatch as the annual chile crop: the local Mexican population. Continued growth in the farming industry has resulted in a marked increase in the number of migrant farmworkers who travel to Hatch from nearby Mexico to work. More and more, immigrants have become naturalized and have begun to establish roots in the area. In Hatch, many old-time residents have found they have new neighbors. Some tell Johnnye they don't like it. "Every time one of us dies off, the house usually sells to a Mexican," she says. "They're here because we're a farming community and the farmers need labor. There's no warring, we're not fighting, but there's definitely two cultures here. They have their way of doing things and we have ours."

Irrigation canals cut across the landscape to feed farm fields throughout Hatch.

Whether real or perceived, this separation of cultures appears to have more to do with conflicting life-styles than with outright racism. The situation has led Johnnye on a cultural crusade of sorts in which she seeks to educate the Mexican population about the values of a small American town. She cites her qualifications to do so: She grew up near El Paso, Texas, in San Elisario, just across the border from Mexico. San Elisario was home to a large Mexican population. She speaks fluent Spanish.

"I understand the Mexican culture, but we have some very narrow-minded people here in Hatch," she says. "I tell them the way we solve the problem is through education. These people need to be educated to live a better life because they've had a hard life in Mexico. They can come up. They want to do better. That's why they're here."

Johnnye's first lesson was on the importance of keeping the community clean. In 1990, she started a "Keep America Beautiful" program that sponsors

Hatch Chief of Police Butch Anderson, below, lives out a policeman's dream each day as he cruises the quiet streets of Hatch.

tree planting, trash pickup, and graffiti cleanup events throughout the year. She has also tried to get more Mexicans involved on local committees, such as the Hatch Chile Festival committee, as well as at the public library and in the garden club, whose members keep the Hatch cemetery weed-free.

On a more controversial note, Johnnye initiated, and the village trustees approved, the hiring of a codes enforcement officer to ensure that residents abide by local zoning laws. The move was made after some newer Hatch residents, most of them Mexican, began crowding camp trailers into areas zoned strictly for residential homes. In a town where the average cost of a two-bedroom home is $87,500 and few rental and low-income options exist, trailers are the only viable alternative for immigrant families. Mexican residents displaced by the zoning enforcement decided to take the village to court. In the lawsuit that has yet to be decided, the Mexicans claimed that, after years of not enforcing local zoning laws, the village had violated their civil rights by beginning to do so now.

"They're saying that we're trying to push the farm laborers out of town, but you can drive up and down these streets and see that they're everywhere," Johnnye says. "We're trying to make the landowners keep this town clean and a safe place to live. We never even used to lock our doors at night; now we lock them all the time. Now, is that any way for residents of the Chile Capital of the World to live?"

There were thirteen burglaries in Hatch in 1995; the yearly average is twelve. In 1994, Hatch had its first homicide in eight years. Drinking, fighting, domestic violence, burglary, and larceny are among the most common crimes in the village, fifteen- to seventeen-year-olds the most common criminals. A few years back, there was a gang of about twenty-five fourth to ninth graders in Hatch who called themselves "Locos 13." Butch Anderson, chief of the Hatch Village Police, coached most of them on the local football team.

"Hatch is a policeman's dream," Anderson says one morning while cruising the village in his shiny blue patrol car. Driving fifteen miles per hour, past the Hatch Public Library and the railroad tracks, past Johnnye Hammett's house, and along a series of streets named for dead presidents, Anderson waves at practically every passerby.

"In terms of crime, there's very little major crime. We've got our share of burglaries, our share of drinking, but as far as violent crimes go, we don't have many. People around here ought to realize how fortunate they are.

They think they've got problems, but the way I see it, they're pretty much immune from the city kind of crime."

Anderson, a casual cop, is a native of Carlsbad, New Mexico, and a thirty-year police veteran. He was based in Hatch as a New Mexico State policeman before retiring from the state force in 1987. In November 1988 he was hired as Hatch's chief of police, a position he held until 1991, when he took a job as Sierra County undersheriff. Two years later, Anderson returned to Hatch to oversee the village's force of four full-time officers.

Lush swathes of farmland border the roadsides in Hatch.

On this day, he is trying to come up with some clues after a few local businesses began posting warnings to parents about LSD-laced "tattoo" stars being found around town. The fliers say the problem is "growing faster than we can warn professionals and parents," but according to Anderson, the drug problem in Hatch isn't nearly as big as residents are making it out to be.

"I'm not going to say we're drug-free; we're like every town, we've got our share," he says. "But it's a little problem, nothing compared to other places I've seen."

More than anything, liquor is this community's drug of choice, and it can be linked to much of its crime. Yet there is not a single bar in Hatch. Throughout the years, bars such as Jack's Tavern, Porky's, the Blue Moon, and the Pioneer all have fallen victim to the conservative leanings of some village residents. They have eventually been replaced by three package liquor stores that do a brisk business. The stores are currently being targeted by residents opposed to the New Mexico State Legislature's ruling to allow Sunday liquor sales. But Anderson suspects the real target is not the liquor industry; as with many of the controversial issues in Hatch, this one may be rooted in the Mexican migrant worker debate.

"One reason the package liquor stores do so well is because local farmers often pay their workers at pay stations in town that are right next to package liquor stores," he says. "So you've got all these workers sitting, waiting at the liquor stores to get paid. It makes some sense. If you've been outside all day with chile sacks on your back, a cold beer is going to sound pretty good."

Studies have shown that crime increases with the full moon. In Hatch, Anderson says, crime follows the course of the chile crop. By

early December, when the chile-picking season is wearing down, the criminals in Hatch, like the land they work, tend to settle in for a winter's rest. December and January are fairly quiet months, although the sudden idleness and lack of income have been known to motivate a burglary or two. In February and March, as the soil is prepared for planting, local criminal activity might surface in the form of domestic violence or drunk driving. When the onion crop starts coming up about mid-May, the pressures and hard work involved in the harvest seem to have a calming effect on the community. The mellow crime mood flows into the green chile harvest, which usually begins in July. During peak picking season, when farmers and laborers work long, ninety- to one-hundred-degree days, crime virtually disappears in Hatch. As the days grow cooler in late September and the chile changes from green to red, work begins to taper off. Crime climbs for a few months until winter settles in the valley again.

"If we ever get this mechanical chile picker going, things are going to change," Anderson says. "Crime could increase because people will have less to do."

For now, however, the chief will begin his daily patrol at 6 A.M. with a drive along the dirt roads that traverse the grain silos and the railroad track. Stepping out of his patrol car, he'll pick up the empty beer cans and the broken bottles of Jack Daniels that the migrants—and the farmers—have left behind. "We've got a lot of good people here, but we do a lot of in-fighting," he says. "From my standpoint, the migrants aren't much of a problem. We have many productive Mexican families in this town who have earned the right to be here. I say, let's live and let live unless somebody's violating your rights."

The house on Wilson Street is painted a screaming shade of yellow. Ruby red chile ristras, in several lengths and shapes, hang on every available wall, fence, and tree. The family who lives there, the Perez clan, has worked all day to cover their home with the ristras, wreaths, and other decorations they have made by hand. At harvesttime, the older family members work in the chile fields alongside other Mexican immigrants. They then go home and transform the literal fruits of their labor into chile decor. While the two youngest children shred cornhusks, their parents and two older siblings string together chiles, three pods at a time, then place the fluffy husks on top.

Jesús Perez, a legal immigrant from Chihuahua, Mexico, has lived in Hatch for nearly twenty years. An extremely friendly man, Perez hesitates to reveal much about his past other than to say that he moved his family to the rental home on Wilson Street in 1987 and that he always begins his workday by 5 A.M. He leaves it to his daughter, Yanet, a pretty, gregarious teen, to tell how their family supplies wholesale chile ristras and other decor to chile businesses in Hatch and beyond.

"They come from Texas, Arizona, Utah, all over, to buy our work," Yanet explains. "Someone from Taos came at 2 A.M. this morning to pick up their chile, that's how bad they wanted it."

Yanet speaks Spanish to her parents as a sign of respect but shows her successful assimilation into her adopted home by switching to English with ease. If not for her strong sense of her native culture, Yanet could pass for most any all-American teen. However, she has learned to look beyond such small-town dreams as being a cheerleader or a candidate for Hatch Chile Queen. As the bilingual representative of the family cottage industry, she puts her energy into securing her family's future.

The Perez family, originally from Mexico, has made the most of the town's chile industry. Opposite: Every fall, family members display and sell the fruits of their labor outside their house.

"There's tension in Hatch about us Mexicans," Yanet says. "People say there's trash in Hatch because of the Mexicans, but I've seen others, white people, throw trash, too. I even have some friends like that. They judge the whole race on the way one person behaves. I tell them, 'Hey, remember me? I'm Mexican, too.'"

Like many teens in Hatch, Yanet hopes to enter New Mexico State University in Las Cruces, forty-five miles south. There, she hopes that her bicultural background will be seen as an asset rather than as a threat.

"I hate to leave my family, but my dad says to do whatever makes me happy," she says. "This is America. I can be anything I want. I don't have to pick chile the rest of my life."

Veronica Valdez, top left, reigned as queen of the 1995 Hatch Chile Festival. The 24th annual festival featured fresh-roasted green chile, red chile ristras, chile-growing and chile-cooking contests, carnival rides, and a parade.

Approaching the Hatch Airport on Labor Day weekend, the first thing one hears is the roar of rows of propane-powered, green chile roasters. As the chiles toss about in the giant rotisseries, which mechanically cook fresh green chile pods and blister their skins for easy peeling, a smoldering smell fills the hot and dusty air. The scent is more aromatic than foul and has a strangely soothing influence on the senses. People crowd around the roasters just to catch a whiff. Eyes closed, they inhale, slowly, deeply, over and over again.

The parking lot at the Twenty-Fourth Annual Hatch Chile Festival is filled with California, Texas, New Mexico, and other miscellaneous state license plates. They have come to this sunbaked corner of town to take in the intoxicating smell of fresh-roasted chile plus two days of chile-related events. Surprisingly, other than the ingredients in the green chile relleno burritos, the sack of green chile that is raffled off every half-hour, and the various vendors selling Hatch chile from the backs of trucks, there is not that much chile at the Hatch Chile Festival. One can register to win a Florida vacation, watch performances by local square dance troupes, take the kids for a pony ride, or test Margo's psychic skills. An "art gallery" filled with the works of local artists features cute and fluffy kitty cats as the predominant theme while fajitas, funnel cakes, and other greasy fare are served out of makeshift booths or the tiny sliding windows of RVs. Right next to the food are the undulating Tilt-a-Whirl and the spinning Swagger, rides that are sure to give someone a case of heartburn—or worse.

At most, the annual chile parade and contests for best chile recipe, wreath, and ristra take up three hours of the two-day event. Still, an estimated ten thousand people a day come to Hatch to see which farmer's Big Jim, Joe E. Parker, or New Mexico-6-4 chiles have been deemed the best of the local crop. The fresh prizewinning pods, their skins polished to a glistening green, are displayed on white paper plates.

In the middle of it all stands Veronica Valdez. Tiny plastic chile ristra earrings dangle from her ears and a rhinestone tiara rests on her head. Hanging diagonally from the shoulder of her blouse to the tip of her skirt is a glittery red-and-white sash that proclaims her "Hatch Chile Queen 1995," winner of a competition open to local high school juniors and seniors. Veronica is the official chile ambassador of the Twenty-Fourth Annual Hatch Chile Festival.

"I guess my job is to let people know we exist," Veronica says between a series of handshakes and photographs. "I represent the Chile Capital of the World."

A native of Hatch, she says she was "ecstatic" when, a month earlier, she was chosen as queen from among four other local girls. She modeled a green gown with sequins around the trim and neck. Then she secured the title with a provocative speech that dared ponder life in Hatch without chile.

"I asked what would Hatch be like if it wasn't the Chile Capital of the World but the Broccoli Capital of the World instead," Veronica says. "Hatch eats chile for breakfast, lunch, and dinner, but broccoli is something we don't eat all the time. This would be a completely different town if we didn't have chile. I mean, how would we adjust?"

On Chile Festival weekend, it is indeed hard to imagine what Hatch or, for that matter, the rest of New Mexico would be like without the annual chile crop. The fiery breath of roadside chile roasters pulses like a heartbeat through the village as people young and old, Americans and Mexicans alike, hawk Hatch chile side by side.

"Get your Hatch chile," a man yells over and over at passersby. "Get it hot, mild, medium, extra-hot."

In a weed-filled field beside the Dairy Queen, just before the road turns out of town, an abandoned festival float sits alone in the dirt. Red-and-green crepe paper chiles hang from its sides while a glittery banner is pasted across its front.

Chile por Vida, the banner says.

Chile for Life.

HATCH
CHILE